Contents

To all who loiter with The LRM, with love and golden apples,
because the streets belong to everyone

The Feminist Art of Walking

'Rose evocatively demonstrates the power of walking as not only a source of individual connection to place, but as a critical form of collective engagement that helps us rediscover, and fight for, what matters to our communities.'

—Leslie Kern, author of *Feminist City: Claiming Space in a Man-Made World*

'A moving and delightful wander through the personal and the political in walking. A feminist manifesto for everyone who has felt the oppression of the city and wants to claim it back as a place of joy.'

—Paula Serafini, Senior Lecturer in Creative and Cultural Industries, Queen Mary University of London

The Feminist Art of Walking

Morag Rose

PLUTO  PRESS

First published 2025 by Pluto Press
New Wing, Somerset House, Strand, London WC2R 1LA
and Pluto Press, Inc.
1930 Village Center Circle, 3-834, Las Vegas, NV 89134

www.plutobooks.com

British Library Cataloguing in Publication Data
A catalogue record for this book is available from the British Library

ISBN   978 0 7453 5099 8   Paperback
ISBN   978 0 7453 5101 8   PDF
ISBN   978 0 7453 5100 1   EPUB

Typeset by Stanford DTP Services, Northampton, England

Simultaneously printed in the United Kingdom and United States of America

EU GPSR Authorised Representative
LOGOS EUROPE, 9 rue Nicolas Poussin, 17000, LA ROCHELLE, France
Email: Contact@logoseurope.eu

First Steps

I am a Loiterer. I am The Loiterers Resistance Movement. Once a month, every month for almost 20 years now, I have stood on my own somewhere in Manchester and loitered. I've put out an invitation for anyone interested to come and explore our city with me. Then I wait to see who will turn up. We will walk together, not following a map but letting ourselves drift. Maybe we will use playing cards or our noses or ears or pigeons to guide us or perhaps we will search for something suggested by an artist many miles away. I don't know where we will go or what we will discover.

I am a loiterer because I want to have conversations with, and about, the streets we live in.

I am a loiterer because I am curious, I want to explore and ask awkward questions.

I am a loiterer because I want to make creative mischief and subversive new maps.

I am a loiterer because I cherish serendipity and the poetry of the pavement.

I am a loiterer because everyone, and everywhere, is interesting and mostly good.

I am a loiterer because there are places I feel scared to go alone.

I am a loiterer because I want to challenge forces trying to divide and segregate us.

I am a loiterer because I like to take theories for a walk, test them out, give them a spin.

I am a loiterer because I collect stories to share with the city I love.

I am a loiterer because once I was lonely, and now I am not, and the people who join me are friends, even if only for the afternoon.

I am a loiterer because I want to play out, be silly, be surprised and open to what's around every corner.

I am a loiterer because I believe public space matters and it should be open, accessible and free for everyone to enjoy.

I am a loiterer because if we don't use these spaces, we could lose something precious.

I am a loiterer because I walk myself into being here and so I have become part of this city. It belongs to me now.

I am a loiterer because I can, and I want to hold space for those who can't yet.

I am a loiterer because I believe walking can change the world.

The first time I loitered I was astonished how many folk joined me. We emerged from an underground bookshop to drift together, an organic, shimmering mass, shifting leadership at every corner. A school? A shoal? A murmuration? It was dark as we wandered down dank alleyways and across litter-strewn carparks – so much empty space, so many carparks. Companionship fed curiosity and made me brave as we followed the towpath into Ancoats canal basin and onto Miles Platting. Together we traced waves of history and regeneration, smelt fireworks, cannabis, chip fat. Someone whispered we were in the watery realm of Jinny Grinteeth, disgusting hag, eco warrior, guardian of the lost or devourer of children depending on the bedtime story your mam chose. Occasionally, we heard snatches of music, drum rolls and guitar squalls, from dilapidated warehouses commandeered as rehearsal rooms and squatted studios. Constellations of red stars blinked above us, each one a crane reinventing Manchester. The Card Room Estate had been demolished and the first fancy new homes were beginning to appear for New Islington pioneers. On subsequent visits we would spot topiary dinosaurs and intricate utility covers memorialising hearts, diamonds, clubs. Designer hubris before the financial crash. Money would run out, work would stall, a community vigil held to save the Victorian Ancoats Dispensary building. Redevelopment would take decades. We couldn't know that then, as we turned around, avoiding the retail park and wandering into Little Italy. A redundant church, placid terraces with neat curtains and garden gnomes, Serafino's stone. It felt eerily deserted, no people, just geese flying overhead and a fleetingly glimpsed fox. A city in flux, dreaming itself into being, a paradox of new window frames and burnt out cars. Eventually, we headed to the pub to hunt for wreckage from a crashed zeppelin and swap travellers' tales. I asked folk to write a few anonymous lines reflecting on their wander:[1]

'the vague but powerful inner pull of something I felt deep in me … darkness and strangers'.

'this is where we took a shortcut that was three times as long and this is where I worked that Summer and we watched a building burn down, and this place is sad and this place is lonely and over there is a big, big, BIG gorilla'.

'Condemned buildings full of asbestos and despair, but dignified in a crumbling and scorched kind of way, buildings half demolished and buildings left naked by their neighbours' destruction. What happened here?'

'Drifting, shifting/altering states of consciousness from artificial purpose to liberated purposelessness. Becoming explorers. Becoming Lost. Touching parts of the city I've never touched. Peering through holes in walls and boarded windows. Shifting perceptions, new playgrounds, old living quarters. Making the city one big Temporary Autonomous Zone'.

The second time we loitered it was a sunny bank holiday afternoon. We tried to decipher the word on the street to see where it would take us. Mixed messages. A cacophony of graffiti, a palimpsest of posters, tags, stencils. Cartoon cats and astronauts, 'Live at The Roadhouse', 'Free Palestine', 'Eat More Pies', 'Happy Hour Every Day', 'Paddington Bear says Migration is Not a Crime'. A small figurine – an imp? A pixie? A fairy full of dust? – was nestled on a ledge above a crumbling wall. Ghost houses left outlines on their still standing neighbours. Echo location. On a Northern Quarter backstreet we were hailed by a man leaning out of a third-floor window, wanting to know who we were and what we were doing. Our mumbles amused him so we were invited into a building mid-transformation. Our host told us he had been in the city just two weeks, came to help his cousins with their work. We followed him up a wooden ramp, past cobwebs and massive weighing scales, broken mirrors and unidentifiable industrial detritus. Men in overalls were building a TV studio and radio station. I listened out for ages afterward but never caught a broadcast. By the third time of loitering I was enthralled and knew I was not alone. Not long afterwards The Loiterers Resistance Movement (LRM) was officially named. Then, in February 2007, it was the first First Sunday. Ever since then I have shared a monthly invitation and waited to see who will join me.

The last time The LRM loitered together was a fortnight ago. We threw dice and looked for portals into the future. On

Shudehill we speculated about Withy Grove Stores, no one can remember this mysterious purveyor of safes and swivel chairs ever opening its doors. We spotted tributes to gargoyles and gangsters, talked about San Francisco, chemotherapy, ice cream, artificial intelligence, the possibility of time travel and our best loved books. We asked ourselves can we imagine a better city, and if so how do we get there? Everything and nothing written with our feet.

Introduction

Beyond the pedestrian

Walking is an everyday, mundane and taken-for-granted activity most of the time. However, it can have extraordinary resonances and be transformed into something wonderful. It can be more than pedestrian – and I want to reclaim that word because actually walking is anything but. It provides an opportunity for multi-sensual exploration and a deep connection with space, place and communities. Walking can be a powerful artistic, creative and political act and I want to celebrate some of the women who are leading the way. Before I show you what I mean it is important to clear up a few common misconceptions first. There is no one Great Walk or way of walking. The number of words we have for moving illustrates the diversity: stomping, strutting, sashaying, skulking all convey quite different meanings.

The walking this book focuses on is for anyone and everyone who wants to give it a go. 'Walking' my way includes wheelchairs, sticks and mobility aids, it welcomes all kinds of bodies and movement at any speed that suits. I acknowledge we live in an environment that does not reflect this belief, and many obstacles are faced. However, my walking is hopeful and believes in a better way. A holistic, generous and open view of access, walking and indeed space itself benefits everyone. An intersectional understanding also considers exclusion around gender, race, class, sexuality, age, body size, faith and other factors, opening up and expanding what we mean when we say we are going for a walk. These themes will recur across the book, with discussions around gender and disability as my main focus. Perhaps this is the start of a conversation and I hope others will continue to take on a critical analysis of other identities as they relate to walking.

Walking can be a creative act. In *Walkscapes* Francesco Careri[1] suggests that walking may have been humanity's first aesthetic act because after basic physical needs are met, our attention turns to exploring and modifying the landscape. My focus is largely on how walking is used by artists as a way to express something, and so it's probably useful here to share some definitions.

'Walking art'

Walking art covers a range of practices that use walking as a medium. For some artists the walk is the thing, a performance in and of itself. For others the walk is a catalyst or inspiration for making further work. How to, and whether to, document or represent the walk afterwards can be a challenge. 'Walking art' is

> [A] creative medium in which the act of going for a walk is the primary material. Works in the medium of Walking Art use varying tactics or approaches and secondary media to engage audiences, such as: audio walks, listening walks, performance walks, guided tours, maps/self-guided walks, instructions and scores. These have connections and contexts that link to creative practices in the visual, textual and performing arts, and many walking artists have arrived in this field from the disciplines of Fine Art, Writing, Theatre and Performance, Dance and Music.
>
> For some creative practitioners the experience of the walk is the primary artefact, for others it is a catalyst or conduit for interpretation or documentation via a range of media.

Walking art activates interconnections and operates as real-world intervention whilst simultaneously fore-grounding aesthetic considerations.[2]

As Francesco Careri suggests, this idea that walking can be art has a long history. However, walking art became regarded as a practice in its own right from the 1960s thanks to artists such as Richard Long and Hamish Fulton. Long's *A Line Made by Walking* (1967) is regarded as pivotal. There is also a relationship to land art, and performance art, where innovators such as Yoko Ono and Stanley Brouwn involved the public in their work. Many contemporary walking artists are explicitly socially engaged and participatory, and this book will explore some examples of what this actually means in practice. I have provided a list of recommended reading if you want to learn more about the histories and philosophies of walking art.

There is also a long relationship between walking, writing and thinking. Rebecca Solnit[3] suggests this is because 'the mind, like the feet, works at about three miles an hour'. Many of us can relate to how refreshing a walk can be when you are trying to think. In *Wanderers: A History of Women Walking*, Kerri Andrews[4] introduces us to women who use walking to inspire and shape their writing. Contemporary authors such as Anita Sethi, Lucy Furlong, Sonia Overall and Ceri Morgan use walking as a catalyst for their work. There is something about walking that can invite introspection, the sparking of new ideas and sometimes also playfulness. Walking art is not just for designated artists, one of the reasons I love it so much is because it blurs the line around who, or what, an artist is. There is also a wider, and overlapping, idea of creative walking which may be familiar to you although you may call it something else.

'Creative walking'

Creative walking is something many people do without labels, or permission, and that is exactly as it should be I think. The term 'creative walking' refers to

> activities that people or groups may undertake whilst walking, which have some kind of imaginative, playful or task-based framework. This could include, for example, looking for rainbow posters or red cars, drawing shapes on maps or hunting for treasure using digital apps or taking photographs to share online. It also includes creative interventions, made by others, that you may encounter on a walk (e.g. chalk messages, art trails, fairy trails, knitted decorations or similar).[5]

We are always in flux when we walk, betwixt and between, our minds and bodies in transition. Walking is also an embodied experience, you cannot detach a walker from their physical body or the environment they walk through. To walk is a subjective and sensual experience, we are open to the elements, textures underfoot, sights, sounds and smells surround us. We do not, cannot, all experience the same walk in the same way even if we are walking side by side because the sensations we feel are uniquely ours. My assertion – that walking, and the streets should be for everyone – is not a reality yet. As a disabled person I am hyper aware of pavement surfaces, slopes, and steps. When there is ice my world shrinks as I fear slipping, my body remembers pain and bares scars. As a woman I am constantly alert to gendered harassment and violence, something I, and every other woman I have known, have experienced. That's not hyperbole, and it will not stop us walking, but we know the journey has never been equal. There is no 'just' walking.[6] We learn through our own bodies and experiences alongside stories shared by our families, friends, communities and the media. They tell us to be careful, to limit where we go and when, because our gender makes us targets. For Black women and women of colour, racism can also impact their experiences of walking. Some women face Islamophobia, antisemitism, homophobia, transphobia and other intersectional oppressions because they are wrongly deemed unworthy of taking up space. Learning from experiences outside my own, and trying to be an ally, also shapes my walking. In this book, we will meet artists, writers and activists all working to make walking safer and more accessible for everyone.

The idea that walking can be a radical, perhaps even revolutionary, act can be traced back to the Avant Garde art movements of the early twentieth century. Movements and groups such as the Surrealists, Lettrists and Dada were all reacting in different ways to the rupture and chaos of the early twentieth century and the horrors of World War I. They radically redefined what art could be. This was taken up by the SI (Situationist International) (1957–72). The SI were radical thinkers, writers, artists and architects who wanted a total revolution.[7] Their de facto leader Guy Debord wrote *The Society of The Spectacle*[8] which discusses how capitalism, and mass media, alienates and corrupts us. The illusory spectacle tempts us, sells us false dreams, stops us connecting with what really matters. This was well before social media, and his influential analysis remains pertinent. Think about all the images we consume today, willingly or not, and how they impact us. Consider billboards, product placements and other advertising as psychic pollution. What about the impact of social media, beauty standards, algorithms provoking reactions, favouring conspiracy theorists or leading into the manosphere. How might these influence the viewer? These ideas remain worth thinking about.

'Psychogeography?'

Another big idea Debord had was psychogeography; although it was not something he or the SI dwelt on, it has had an incredible afterlife. Debord defined psychogeography as 'The study of the precise laws and specific effects of the geographical environment, consciously organised or not, on the emotions and behaviours of individuals.'[9] That study was to be conducted by walking through the city, being attentive to changing atmospheres and energies. It was a reconnaissance mission to understand how capitalism shapes the territory of the city, but it was also an attempt to resist and disrupt those processes too. Psychogeographers use the dérive or drift to meander through space rather than taking a utilitarian approach. Greil Marcus said, 'The point was to encounter the unknown as a facet of the known, astonishment on the terrain of boredom ... the physical town replaced by an imaginary city.'[10] This gets at the heart of psychogeography's lasting appeal. The dérive disorients and opens up the wanderer's imagination. That imagination is a new, boundless place, unprescribed. Psychogeography was described by Debord as 'pleasingly vague', which means it is adaptable to any time and place. Psychogeography uses the body as a tool for exploration, and encourages a deeply attentive engagement with space. Its playfulness transforms the pedestrian into something truly extraordinary. This continues to hold today, as we shall see later.

However, the SI were problematic in many ways and their legacy should be contextualised. The sheer number of members of the SI (70+ but never more than a handful at any one time) demonstrates the factionalism and constant expulsions, falling outs and quarrels. This feels familiar to anyone on the left who may wonder where the revolution is if we cannot even agree how to tie our shoelaces. However, it hints at more insidious issues. Much Situationist writing – and that which followed in its wake – has a distasteful neocolonial tone to it.[11] There is a frequent tendency for the SI to view themselves as conquering and penetrating space, discovering somewhere for the first time. This demonstrates scant regard for those who already inhabit it, disrespecting or disregarding the experiences of others. Debord could be very dismissive of women's experiences, for example when he maps women's movements across Paris and he finds they travel less far and less freely than men. There were women involved with the SI, most notably Michèle Bernstein and Jacqueline de Jong, but their contribution has historically been minimised.[12]

Considering these issues, it is worth pondering whether there is any value in drawing from the psychogeographical archive. Personally, I had such an epiphany when I first encountered the idea of psychogeography I resolved to subvert it and become an anarcho-flâneuse. It is probably instructive that I had not actually read much of the literature at this time. If I had, I may have been put off because many of the SI texts are somewhat boring and arcane to read. For a movement seeking the 'revolution of everyday life' they frequently forgot to engage meaningfully with their intended audience. Significantly, because I learnt first through the filter of activist friends, rather than the original texts, I had not had a chance to imbibe the message that this kind of walking wasn't for me. Nobody told me that I – a queer, disabled, working-class woman from an English suburb – could not be a psychogeographer, so I just went and did it.

In the years since the SI, psychogeography has mutated, and evolved, in various ways. The most prolific is as a literary tradition. This is probably what most people think of in terms of psychogeography (if they ever think of it of all, that is). Iain Sinclair and Will Self are the most famous of those to use walking as a catalyst for investigating place and

7

its many tangents. Others include the deep topographer Nick Papadimitriou, author Rachel Lichtenstein and filmmaker Patrick Keiller. Each in their own way makes connections in their landscape and renders visible the construction of place. Psychogeography has also influenced wider culture. The SI gave very, very good slogans. They impacted rebellious DIY cultures like punk, although it is not always clear how aware participants were.[13] The SI's ludic tendencies and urge to create spectacle can be seen in protest movements such as Occupy and Reclaim the Streets. The Situationist's legacy is perhaps most overt in loose collectives that emerged in the 1980s and 1990s onwards such as the Association of Autonomous Astronauts, Proflux and Manchester Area Psychogeographic.[14]

Many contemporary walking artists, including several of those in this book, also draw on psychogeography. Tina Richardson[15] provides an excellent overview of what this means in practice in the UK today. Tina identifies qualities that she thinks define a 'new psychogeography' which is heterogeneous, critical, strategic and somatic. It is vital to me that psychogeography is always a doing word, a verb as well as a noun. *Doing* psychogeography can feel like you are joining invisible dots in a landscape, nothing is 'natural'. That critically engaged and political edge is what I think makes psychogeography such a useful tool. I believe all psychogeographical work includes an element of walking art or walking as aesthetic practice. However, not all walking art is psychogeographical, and the label psychogeographer is actively resisted or rejected by some. I don't want to dwell on semantics, because regardless of label, too much of the most interesting walking work is not sufficiently celebrated.

Banishing 'The Flâneur'

There is often an assumption, reinforced by many books and events, that walking art and psychogeography are just for men.

The stereotype of an ideal walker in general has long been male, as Lauren Elkin[16] suggests, 'as if a penis were a requisite walking appendage, like a cane'. This in turn reflects wider experiences of public space as a traditionally male domain. Women can feel the pressure of this structural misogyny when they, we, walk. The flâneur embodies the myth of the perfect walker. He is male, wealthy and able-bodied, he can walk freely anywhere he chooses. Problems arise because his body is privileged above all others and viewed as universal.

Rebecca Solnit is clear he never existed. She discusses the very different implications of the phrase 'street walking' when applied to men and women and gives many examples of cultural norms and legal prohibitions intended to limit women's movements. There remains debate about the possibility of the flâneuse because of the male gaze. Lauren Elkin writes powerfully about her own experiences, while Griselda Pollock,[17] Janet Wolff[18] and Helen Scalway[19] are amongst those who discuss the limits placed on women walking. Elizabeth Wilson's book the *Sphinx in the City*[20] provides a historical overview of women within the city. She documents a culture seeking to control women's access to public space, influenced by a fear of female sexuality and patriarchal norms seeking to limit the autonomy and freedom of women. This in itself made women's very presence in cities a problem. In *Feminist City*, Leslie Kern[21] discusses how misogyny is built into the fabric of our landscapes. Chapter 3 'Eastbourne' and Chapter 6 'Liverpool' explore what this means for women walking. In the UK in recent years, more attention has been turned to the safety of women, notably since the murder of Sarah Everard by a serving police officer. Awareness raising has not yet translated to meaningful large-scale action.

These threats do not stop women walking, they never have. Recently, much energy has been focused on reclaiming space for women in walking, walking art and psychogeography.

Much of this work has been done by women who are themselves also walking artists. 'Woman' is not, never is, a genre so
part of the labour has been revealing the diversity, richness and
depth of women's walking work. Alison Lloyd's[22] research
provided valuable historical analysis which situates the work
of women artists between the 1960s and early 1980s. She discusses how conceptual artists including Marie Yates, Michelle
Stuart and Nancy Holt used walking as part of their process.
Alison herself was a hillwalker as well as a photographer and
fine artist; her work is discussed in Chapter 5 'Sheffield'.

Dee Heddon and Cathy Turner[23] wrote what are widely
acknowledged as the first scholarly papers on women walking
artists (their own art is discussed in Chapter 4 'Stockport,
Ashton-under-Lyne and Glossop' and Chapter 6 'Liverpool').
Dee said they had 'a feminist motive, attending to work that is
there that is invisible … to make it visible to underscore and
centre its importance'. Jo Norcup[24] has also been doing this
work through radio shows aiming to bring women walking
artists to a wider audience. Her first broadcasts for Resonance
FM were a direct reaction to programmes which only featured
male voices, or where Jo felt women were spoken over and
belittled by men. This catalysed her 'I thought if I don't step
up and do this who will … I wanted to create (this) possibility of space … it wasn't that women artists hadn't walked
but there was this mediated control.' Jo knew the work was
happening and created the space to celebrate it. She told me
when she hears about the flâneur: 'it makes my toes cringe, I'm
waiting to hear the same tropes … it's boring and dull, we've
done this already'. I absolutely agree with Jo that the flâneur
is boring, and I don't intend to give him any more attention.
If I am honest, I would rather not have involved him at all,
but I couldn't face the inevitable assumption from the usual
suspects that I was unaware of history.

Around the same time as Jo was making radio, Clare Qualmann and Amy Sharrocks produced the *Study Guide for Women Walking*.[25] This is a document cataloguing *Walking Women* events held in London and Edinburgh in 2017. To read the guide is to want to dive in and go for a walk. Highlights include *An Intimate Tour of Breasts* with Claire Collison, described as 'a three-hour walk taking in high street and high art, tea rooms and fitting rooms. Participants will be exploring the mythologies and commodification of breasts throughout history to the present day'. Idit Nathan invites you to join her *Taking a Die for a Walk* and Barbara Lounder explores the legacy of the Halifax explosion on communities, especially Indigenous people, in her hometown in Nova Scotia. These walks are ephemeral, gone, but documents remind us they were real and their memory resonates. Clare Qualmann is also a founder of The Walking Artists Network (her own walking art is discussed in Chapter 3 'Eastbourne' and Chapter 4 'Stockport, Ashton-under-Lyne and Glossop'). She is clear that not only do we need to create new work but also champion others because

> every sphere is clouded by the patriarchy and it is lots of work to unpick, untangle and relocate walking art … writing about art if we think it is important, is activism. Write an article, blog, zine, Wikipedia article, share it, make it public so it becomes on the record. That is art history. We have a responsibility.

This is why my primary focus in this book is on women, explicitly including Black women and women of colour, trans, non-binary and genderfluid women and women of all ages, sexualities and body types. There has never been just one walking woman. We are many and multitudes.

I am a loiterer

So, what do all these terms mean for me, and what do they have to do with me hanging around Manchester seeing who will turn up? I am a loiterer. In 2006[26] I co-founded The Loiterers Resistance Movement (The LRM), a psychogeographical collective based in Manchester, North West England. At the time, I was part of the volunteer collective at The Basement, an anarchist social centre. We aimed to provide a haven from the commercial pressures of the city, a free space for collaboration, conversation and alternative visions to cross-pollinate and flourish. The Basement was an incubator for a range of social justice campaigns that shared a broad anti-capitalist ethos, including environmentalism, animal rights, refugee and asylum seeker support, open source software, LGBTQIA+ liberation and others. The majority of the founders had been involved in squatted 'Okasional Cafes' and wanted a more secure base that would enable wider participation in their vision. They secured a legally rented spot in the Northern Quarter, then established as one of the creative areas of the city. I got involved just as a musty cellar was being renovated into a community resource which included a radical bookshop and library, exhibition space, meeting room, free internet/computer spaces and a vegan cafe.[27]

Given my singular lack of practical building skills, I became mainly involved in paperwork (yes, anarchy needs admin and washing up done too). I also spent a lot of time working front of house in the cafe and bookshop, listening, learning and absorbing ideas emerging from the milieu. This is where I first heard about psychogeography and the SI, in an atmosphere charged with experimental ideas and a 'give it a go' DIY punk energy. We didn't have, or need, much money but we were all passionately doing our bit to make our city better and more interesting. We were committed to building a culture that offered an alternative, outside the consumer economy. In this spirit, all LRM events have always been free and open to anyone, because if we believe these streets belong to everyone, and we really want to pay attention to all their stories, how dare we sell tickets? We loiter in the spirit of mutual aid and exchange. The Basement closed in 2007 after a fire in an adjacent building (see Chapter 4 'Stockport, Ashton-under-Lyne and Glossop'). Our premises was damaged and evacuated, remedial works took years. There was some serious searching for a new home but we couldn't find anything suitable or affordable. There was grief at the closure but a very real sense that the spirit of The Basement was actually in the people, projects and ideas it nurtured and generated.

Parallel to The Basement I was working (broadly) in community development/voluntary sector support across Greater Manchester. Through this I met many other people working in different ways to improve their communities, feeding, housing, supporting, caring, providing services and opportunities. I felt the two parts of my life were frustratingly disconnected. My relationship with Manchester was also changing. I was feeling both enraptured and repulsed. I was falling in love with the streets while at the same time it seemed those very streets were being taken away from me. There was an atmosphere of securitisation, gentrification, enclosure, all wrapped up in New Labour rhetoric. These were supposedly boom years but I could see inequality becoming entrenched. I could also see the far right seeking to establish itself in suburbs as Manchester's wealth accumulated around a shiny core. Naively, I thought walking together could be an interesting and valuable way to start conversations across some of the divides. I wanted to understand the city better, and to have some fun. Frankly, I was also exhausted and at risk of burnout.

Following a series of encouraging conversations in The Basement a few of us organised the dérive which I shared in 'First Steps'. I made a flyer which said:

Winter Solstice Shenanigans

Curious? Please join us for a multi dimensional psychogeographical exploration of Manchester. We'll be making a spectacle of ourselves, finding magick in the mancunian rain, listening to cracks in the pavement and probably getting a wee bit lost. We like to make our own maps not follow them.

Gather at The Basement, 6pm December 21st 2005. Wrap up warm, we'll be off on an expedition.

There were no names, but the flyer included my phone number, and an email address belonging to Alex Bridger,[28] one of the earliest loiterers and co-organiser of that first event. That first invitation included a Venn diagram I drew to try and explain what I thought psychogeography was. I still use the image sometimes, although I wish it included walking within it. The intersecting circles are labelled art, social history, serendipity, de/regeneration, map making, situationism[29] and art, and an arrow points to the middle saying, 'psychogeography is here, sort of'. Energised by that first wander, several more followed, usually on equinoxes or other dates we deemed significant. We didn't have a name or particularly coherent message but all shared a critique of what was happening in Manchester and a joy in discovering stories and places that were new to us.

Late in 2006, we adopted The LRM (Loiterers Resistance Movement) name. Loitering felt integral to our ethos, representing a conscious slowing down, a challenge to the productivity of the rational city. We were aware of how simply hanging around has been criminalised and wanted to challenge the demonisation of non-productive uses of urban space. We felt we needed a name because we were organising *The First Accidental International Psychogeography Festival*. It featured a programme of walks, talks, games, DIY map making and an exhibition of Situationist-inspired art. We wanted a collective moniker to protect individuals and make it clear anyone was welcome. When the exhibition ended, many people said they would like to continue wandering together, so First Sundays were instigated. This has meant that once a month, almost always on a First Sunday, The LRM go for a dérive together. I believe this consistency is one of the reasons for our longevity, as it has enabled a peripatetic floating community to form – but back then I never dreamt The LRM would drift on for years.

Nothing has stopped our flow. During the Covid-19 pandemic, we carried on walking together alone,[30] connecting via technology wherever people were, whether they wanted to wander inside or out. Over the years, the membership, or rather attendance, is fluid. Co-founders and organisers have come and gone. There were no SI-style tiffs or expulsions, just changing demands on everyone's time. For multiple reasons I'm still here. So, I put out an invitation and I wait for people, I never know who will come, but they do. I love the ebb and flow, sometimes old friends return after years, and most months will see new faces. It's very rare only to encounter someone once. The tenacity and rhythmic cycle help other loiterers know we are here.

Back in those early days, I wrote a manifesto that has appeared on various postcards, flyers and zines since. It's been edited a bit for clarity but is substantially the same today:

The LRM (Loiterers Resistance Movement) is a Manchester based collective of artists, activists and urban wanderers interested in psychogeography, public space and the hidden stories of the city.

We can't agree on what psychogeography means but we all like plants growing out of the side of buildings, looking at things from new angles, radical history, drinking tea and getting lost; having fun and feeling like a tourist in your home town.

Gentrification, advertising and blandness make us sad. We believe there is magick in the mancunian rain.

Our city is wonderful and made for more than shopping. The streets belong to everyone and we want to reclaim them for play and revolutionary fun.... .

On the first Sunday of every month we go for a wander of some sort and we also organise occasional festivals, exhibitions, shows, spectacles, silliness and other random shenanigans. Please join us, everyone is welcome. Our events are free and open to all: these are our streets and they are yours too.

Why do people come and loiter with me? The reasons are myriad. I know for some it is less about spatial politics and more about conviviality, for others curiosity, but for all of us in some way I think it is connection with people and with place. When we walk together communication feels easier, different – and actually if you don't want to talk it's easy to avoid conversation and tune into other senses. The simple act of being in space changes its use and leaves an impact. We need to keep playing out on the streets or we risk losing them. We animate cities by our actions. In the UK, there is a creeping privatisation and erosion of common land, there is also a culture of constructive use and doing: everything must be productive. Both of these trends must be resisted. The street is one of the few places all can (or should be able to) access and use for free. We encounter difference here and learn to share. There is a need for tolerance and understanding to make what Jane Jacobs calls the 'ballet of the sidewalk'[31] largely a joyous and positive experience. Of course there are people who disagree and disrupt; gendered harassment will be discussed in Chapter 3 'Eastbourne'. But there are also moments of serendipity and delight and a whole world of unmapped possibilities, so please join me for a walk.

A note on sources

The LRM archives form part of the basis for this book. Those archives include the various DIY zines we made, and the photographs many have taken when they have joined us. Mostly, however, those archives reside in us, and the memories that we hold. It should be clear, then, but to remove any doubt, these are my views on our wanders. I hope other loiterers will want to contradict, expand or embellish my versions.

I also draw on evidence from two research projects. The first was my thesis *Women Walking Manchester: Desire Lines through the Original Modern City*[32] which included walking interviews with 43 women. They ranged in age from 20 to 60+ and had a wide variety of life experiences. We met in Piccadilly Gardens and I asked them to show me their Manchester which they very generously did. Some of their stories are included in this work. They all spoke to me anonymously and so are given pseudonyms to respect that. I have indicated a name is a pseudonym by the use of quote marks. Their tales were all different, reflecting diverse backgrounds, but despite this they shared many commonalities. There was an almost universal desire for more green space and concern about something indiscernible, described by some as a soul, that was being lost in the city. Many of the interviews were interrupted by men and this often meant that our conversation changed direction too.[33] The impact of perceived gender was absolutely central to participants' experiences of walking through Manchester. It modified when and where they walked and influenced any precautions they took.

The second is *Walking Publics/Walking Arts: Walking, Wellbeing and Community during Covid-19*,[34] which is also known as WalkCreate and will be referred to as WalkCreate in this book. This project was led by Dee Heddon and co-investigators were Maggie O'Neill, Clare Qualmann, Harry Wilson and myself. This project explored how the Covid-19 pandemic impacted walking, and how creative walking 'could be used to mitigate social isolation and anxiety, maintain health and wellbeing, enhance social connectivity, and facilitate cultural empowerment'. There were several elements to this project, including a survey open to the general public and one specifically aimed at artists. Supplementary interviews were conducted with people involved in walking organisations or activism and with a range of artists. WalkCreate also commissioned eight new pieces of walking art from socially engaged artists. We compiled a digital gallery of artwork created during the Covid-19 pandemic with an open submission process. Finally, we produced *The Walkbook: Recipes for Walking & Wellbeing*. This includes scripts for walks from 30 artists who were addressing some of the most common barriers to walking. All the published WalkCreate works discussed are available free under a creative commons licence. I have occasionally used quotes from interviews with named individuals or anonymous survey responses that have not yet been published elsewhere.

I'll also introduce you to some of my favourite walking artists. I spoke to, or corresponded with, a number of women walkers, artists and thinkers specifically for this book and I thank them all for being so generous in sharing their time and expertise. These women are friends, colleagues and people whose work has influenced my own. Thank you to Alisa Oleva, Anna Minton, Cathy Turner, Clare Qualmann, Dee Heddon, Elspeth 'Billie' Penfold, Helen Stratford, Jane Samuels, Jo Norcup, Julie Campbell, Kiera Chapman, Nadia

Shaikh, Saffron Defiance Swansborough, Sarah Benjamins and Sonia Overall. Some of their interview quotes have been edited for clarity and brevity.

I also draw on the work of many other artists whose work has inspired me. Where possible the notes include links to their work, and I hope you will put down this book and follow them when you can. I have foregrounded the work of women and non-binary artists to better fulfil my aims of reorientating walking art. Men are not banished or dismissed, and they are included where I feel they are relevant and in congruence with the themes I am covering. Some of the folk I discuss may not identify as feminists or agree with my use of the term. I'm not dwelling on the definition here, but I hope my work signals an inclusive and expansive orientation. My feminism is intersectional, transinclusive and always learning, it strives to be better and believes in a more equitable and just world for everyone. The patriarchy harms us all.

This is not an encyclopaedia of walking art and artists, and I know I will have missed some amazing work. It is personal, partial and, like my walks, largely focused on the UK. Even to list those I know I don't have room for risks creating a liturgy and further marginalising those whose work I have not yet encountered. This is a rich, beautiful and kaleidoscopic field of practice and I can only offer a taste. Those I have chosen are here because their work has moved me in various ways.

It's also probably worth mentioning some of the other ways of walking I don't have time to describe here. This is not about processions or parades, although these can be really significant. I love a good carnival and some of the artists I discuss do evoke the carnival spirit. Neither is this a book about pilgrimage, although some artists may view their work in this way. It's also not about protest marches, when footsteps shout loudly to demand change – although there is some overlap, particularly when we consider movements to make the streets safer. I'm not thinking about guided tours, heritage or tourist trails which share a specific story of a place, but some artists are subverting those tropes quite delightfully. I won't be discussing hiking or leisure walking per se, but these walks are entertaining and most are fun. I am mostly focusing on urban, city walks, because that's where I love most, but we will take some trips out into the countryside. The walks in this book are not feats of endurance or heroism. Neither are they about conquest; that is the antithesis of what I want to share with you. I'm not dwelling on how good walking can be for you, because sometimes it isn't, and there is lots of other evidence about the physical, mental health, wellbeing and environmental benefits of walking. We also need to be mindful that there are many people for whom walking is a desperate necessity to seek refuge, safety or freedom. Globally, millions of people walk to flee war, violence and the impact of the climate crisis, and to try to find a better life. Their walks should not be forgotten. Neither should the labour of walking be neglected, it too can be part of work.[35]

Walking is always intrinsically linked with place: we must always walk through and in somewhere. Therefore, I have organised this book around some of the places I know and love best. At the beginning of each chapter I have included a prompt, or script, which can be used if you would like to experiment with your own creative walking. There are six chapters and a conclusion which offers a manifesto of sorts. If you have a die, maybe give it a roll and, in the spirit of The LRM, let that decide which chapter to read first.

1
Manchester

Draw a heart on a map. Follow the line, try to stay true. How do you feel as you walk? What can you see? Who is missing? Can the city touch you? Will you fall in love on the streets?[1]

I moved here in 1999, into a large shared house in Whalley Range. It was a co-operative and later became social housing, we lived in a large Georgian terrace with beautiful cornices and crumbling walls. A decayed splendour, and constant battle against cold and damp. The space was a luxury and so was the cheap rent: it afforded me time and space to experiment even when it was raining indoors. Over the years, housemates included social workers, carers, writers, researchers, art therapists, teaching assistants, community workers, yoga teachers, musicians, planners, students and more. It wasn't always easy, and we didn't all share the same values – no one (including, maybe, especially me) was a perfect housemate. But we all found a sanctuary and an incubator, a place to grow, and there were frequent buses to take us into what we called simply 'town'.

This city centre is where I feel most myself; I get a thrill simply from being immersed in the whirl, savouring familiarity and novelty. However, my favourite places within it are actually where there are a few moments respite from the buzz. Here they are called ginnels, I grew up with twitterns, other dialects say jigger, passage, wynd, snicket, back, alleyway. There's a lovely network from Lincoln Square to St Anne's

Square, folklore says it was a path for taking livestock to market. I've always fancied leading a parade of pantomime cows to claim that ancient right. Go down the edge of The Hidden Gem Church (if it's open, pop in to see Norman Adams' *Fourteen Stations of The Cross* which leave this non-believer moved) and find a narrow path that's a tribute to John Dalton. Chemical symbols decorate cross beams and entry plates. Cross over Bridge Street (mind the cars) and then head between the two plaster masks. There used to be metal umbrellas overhead here and I'm sorry no one seems to know where they are now. They were also apparently a tribute to Dalton, although a rather more oblique one. On King Street there's a very mini arcade, pass silk scarves and designer glasses, and then suddenly you are on St Anne's Square.

There's another cut through – it's a bit too grand to be a ginnel – that was my most beloved place in Manchester. Library Walk, a gorgeously sensuous curve between the Library and Town Hall extension. A world-renowned design, it's still there, but now it's got a building at one end and gates at the other. Many came together to try and stop this; the fight by The Friends of Library Walk went on for years.[2] Hundreds of supporters said this place mattered to them for its sublime spatial poetry – but Manchester City Council said it didn't really exist. They claimed it was meaningless, just the space between buildings. Worse, they said it was ugly, smelly, dangerous for women and ignored our dissenting voices. I still believe our dossier of evidence was the stronger argument, but as the landscape tells us, we lost this one. Despite defeat, the campaign perversely felt strangely successful. Many of us learnt lessons about planning inquiries, council meetings and community organising. New solidarities and alliances were formed which went on to blossom in unexpected ways. Personally, Library Walk instilled in me more confidence, passion and resolve. The campaign introduced me to friends who shared my love of Library Walk and a sense that public space is vital to our city and our sense of belonging in it. We made the power lines, policies and process which govern those civic spaces temporarily visible. Actions have consequences and without Library Walk we could not have saved the towpath at Ralli Quays (see the concluding chapter). I still cringe when I see the carbuncle, but one day I know it will fall. In the meantime nothing can dim my love of the library. Head upstairs to the reading room, absorb the scents of wood, paper and scholarship. Drink in by osmosis the collective wisdom.

* * *

My favourite Manchester writer, and someone whose influence runs through my work, is Doreen Massey. The idea that walks with me now is her conception of space as dynamic, fluid, and dependent on a network of relationships. Massey is clear that a 'place' is more than a line on a map; it is a nexus of entwined stories, each of which is deeper and richer than any cartographic construction. Massey sees places as being constituted by constellations of personal relationships and journeys which create complex networks beyond traditional borders. In her 1994 essay, 'A Global Sense of Place'[3], Massey demonstrates what this looks like in Kilburn High Road where people, policy, goods and services merge from across the world. I can demonstrate it here too at my desk. My computer was made by a Japanese company, with, I assume, a complex and at least partially exploitative supply chain. I'm trying not to be distracted by my emails and messages from friends around the corner and across the globe. My coffee is Ethiopian and is in a mug adorned with the logo of a Scottish indie band, the radio is playing an American singer and I realise I have no idea what kind of wood this desk is made of or who designed my chair. This is just a tiny corner of a single room, in one house, on a street, in North West England. My feet are on solid ground but invisible ties bind me to other people around the planet.

What I love most about Massey's work is her belief that 'Space is always under construction ... It is never finished, never closed. Perhaps we could imagine space as a simultaneity of stories-so-far.'[4] This suggests cities, including Manchester, are always being made and remade and we collectively have the power to change at least our small role in the narrative. Massey does not mean places do not have distinct identities but she offers a generous and outward-looking account of their construction because 'what makes somewhere special is not its internalised history ... but the fact that it is constructed out of a particular constellation of social relations, meeting and weaving together at a particular locus'.[5]

The locus manifests as existing material conditions, however, Massey's conception means they have the potential to evolve. She raises issues of social responsibility and global inequalities, consistently asks questions about power and justice. Those constellations may not be visible to us but any consideration of them must be rooted in material concerns. She warns against becoming lost in the abstract, because despite fantastical visions of world cities 'much of life for many people ... consists of waiting in a bus shelter with your shopping for a bus that never comes'.[6]

There is something beautiful and gently transformative about opening up to exploration, to paying close attention. To accepting the plurality of space and the realisation that the paths we weave through space are just one thread in a tapestry. They contribute a shade we can't really comprehend at the time because we are so focused on where we are and how to avoid tripping up. Emphasising our global connections helps refute racism, xenophobia and exceptionalism. It also provides a challenge to the dominant Manchester narratives. Basically, the story goes: the Industrial Revolution, great men, big chimneys. Then, decline, decay, Factory Records, Madchester, and a successful neoliberal model of regeneration. Rah, rah, rah. This, of course, obscures much.

I wonder what Massey thought of Piccadilly Gardens.[7] This is perhaps the most important public space in the city, and is most certainly the most controversial. Popular consensus (or rather the dominant views on social media) say the current version of Piccadilly Gardens is an abject failure. Many contradictory identities are embodied here. It is a central hub in Manchester and yet also strangely liminal, a space of transit and sometimes fluid boundaries. For example, it is hard to pinpoint exactly where the gardens end on the west side because grass becomes concrete and buskers gather. Piccadilly Gardens is a transitory space – acting as a transport hub and a pedestrian shortcut – but also a destination with restaurants, cafes and bars. It is a workplace with a commercial function, yet allows visitors to devise their own mildly subversive, uncommodified leisure pursuits while sitting on the grass or watching the fountains. The space appears anodyne but political rallies, religious meetings and personal encounters are frequent. There is a small but well-used children's play area, featuring a 'Dilly the Snail' slide. Despite the proximity of a portable police station and CCTV cameras, there are often people blatantly buying, selling and taking drugs, most obviously cannabis and more recently spice (a powerful synthetic cannabinoid).

Piccadilly Gardens is both spectacular and everyday; being a site of festivals, protests and special events as well as mundane practices. Physically, Piccadilly Gardens has undergone many transformations, most recently in 2002 when it was remodelled following an international design competition. This manifestation was initially hailed a success by statutory bodies and the architectural press; a key part of Manchester's post-IRA bomb regeneration narrative. However, there have been criticisms of the design, especially regarding the appropriateness of Tadao Ando's pavilion, lack of regular maintenance and the encroachment of commercial buildings onto the gardens. A popular myth says the architect never actually visited the site and so didn't appreciate how minimalist grey concrete would work in the Mancunian climate. At a micro level, Piccadilly Gardens illustrates many of the key debates around neoliberal development issues, consumerism and control of the city. There are no gates or overt surveillance; ostensibly the gardens are open to all. However, they are regularly patrolled and 'transgressive' behaviours such as street drinking, unlicensed trading or fighting are often dealt with swiftly by security. Thus, the contested boundaries between private/ public, freedom/control, diversity/homogeneity and risk/ sanitisation can all be explored here. Piccadilly Gardens has it all; it is a conduit illuminating the contradictions and complexity of contemporary public space.

I'm actually rather fond of the place because despite all its problems, it is one of the few places in the city centre with grass, and benches, and it attracts all sorts. To prove a point, I once spent a whole day there for *The Bench Project*, talking to anyone who came and sat down next to me. I wish I had stayed for the whole 24 hours I had initially intended but my body was less willing than my mind. I had some great conversations about many things from the weather, to fashionable dogs, the tastiest pasties and the importance of rainbow flags. Earlier this year, I took a group of Sheffield students on a tour of Manchester and we stood on the grass in the middle of Piccadilly Gardens. I asked them for their impressions and they used words like 'vibrant', 'cosmopolitan' and 'buzzing'. Fresh eyes and no preconceptions gave a refreshingly positive perspective, although I suspect the spring sunshine helped.

I met most of the women I walked with for my thesis in Piccadilly Gardens because it's the best conversation starter, everyone has an opinion. These were generally very negative. 'Patti' was scathing:

> I hate it here, I avoid it if I possibly can … it feels ill considered and a bit desperate. I feel embarrassed by it as somebody who has lived in Manchester my whole life. I don't like being here as a woman on my own, because of the way it's been put together, it is now policed severely as well which adds an extra layer of dystopian horror to the place.

'Patti' stressed she wasn't anti-modernism, and shared some Brutalism she adored, but stressed this was a bad example in the wrong place. However, like other women I spoke to who remembered the past, 'Patti' was sceptical and un-nostalgic about its earlier incarnations. Several criticised the sunken gardens which the current iteration replaced. They recalled them as dark, squalid and intimidating.

'Daisy' also dismissed nostalgia endemic on social media. She said, 'I remember it as a garden which was really shabby and nothing like the pictures everyone shares on Facebook about how beautiful it used to look in Victorian times.' That online nostalgia feeds into conservative ideas about architecture and comments sometimes tinged with racism, stereotypes of the working class, and an exclusionary vision of who should access the gardens. 'Daisy' was very angry about the loss in size of the Gardens, the 1992 renovation was in part funded by the sale of land to build an office block on the edge:

> I've never got over my anger about that, I'm pointing at One Piccadilly, that stupid building that they put there … the biggest open space that there was, they just had to chop the corner off and it really pisses me off. This is another example of de facto privatisation of public space.

For many, Piccadilly Gardens holds fond personal memories even if they dislike the architecture. It has been and still is an important meeting place. 'Daisy' works with refugees and asylum seekers, she says, 'it's become a hub for recently arrived migrants and so that gives it a sort of energy and vibrancy that you don't really get … in other public spaces in Manchester so I'm always quite curious to find out what's going on here when I come through'. 'Daisy' highlights that people use the Gardens despite the architecture. There's a natural subversiveness about where, and how, folk choose to congregate, sit or move regardless of the architects' intentions. Several interviewees shared a lot of love for the fountains (when they are working).

Since conducting my interviews, Manchester City Council have announced new plans to revitalise the Gardens. Parts of the wall, and the roof between pavilion buildings, have already been removed, and so has the bench I sat on for *The Bench Project*. The results of the public consultation haven't been shared but computer-generated images released to the press appear to show mature trees and giant screens, presumably digital advertising hoardings. It seems probable the unrestricted space will be further reduced.

Helen Stratford is an architect and artist whose recent work is discussed in Chapter 6 'Liverpool'. In 2008, she worked with Diana Wesser as Urban (Co)Laboratory and together they made a short film, *Manchester Blind Spots* on location in Piccadilly Gardens. Her interest is in how small actions can change a space, and she remembers this as one of her first works exploring what that meant:

I remember shaking when [Diana] sat down on the catwalk (a path across The Gardens) … I was really anxious about that simple action of sitting in that space. I was nervous about what people would do. This was not only because of the heightened security: the – now ubiquitous – street wardens combined with a high community police presence in the gardens, but also because of the official and unofficial encounters that we'd had with people during our on-site research. While asking people questions, one street warden asked us if we had a license to do so, at the same time men propositioned us just for hanging around on the, rather bizarrely but somewhat appropriately, named 'catwalk,' ... Then suddenly a young woman came and sat down next to Diana, she seemed to be directly responding to this active invitation to occupy that space differently.

Later that day, I joined Helen and Diana and danced in the fountains. I remember a joyful and liberating time, and as others did the same the exuberance rippled across the Gardens. I asked Helen what her thoughts about Piccadilly Gardens are now. She felt that, while there were inherent issues in the ways the Gardens had been designed, the architecture had also become a scapegoat, a conduit for criticism about the use of the Gardens. This neglected how it figured in the city's wider imaginary. She also noted a distinct lack of ongoing care and maintenance.

Community arts organisation Get It Done collaborated with Central Manchester Foodbank on *Placemaking Piccadilly*, a project which asked people to imagine the future of Piccadilly Gardens. Their vision is for equality in public space for everyone, and they actively included homeless people and foodbank users. Their 2022 manifesto[8] calls for a careful, caring and community-led approach to the Gardens, with their key ask being for 'the Council, Piccadilly architects, cultural institutions, and key services to commit institutional support and capacity on a long-term basis to developing a community-led, participatory "platform" in Piccadilly Gardens, with spaces, installations, programmes, and services specifically set aside for, and co-designed with, grassroots organisations, charities, and community groups throughout Manchester'.

Their inclusion of people who are homeless and users of support services was important because beyond Piccadilly Gardens, Manchester exhibits many traits that demolish the myth both of a radical city and of welcoming public space. This includes benches designed so they are uncomfortable to sleep on and barriers preventing access to railway arches that had been used for shelter. The entire city centre is covered by a PSPO (Public Space Protection Order)[9] which explicitly prohibits tents, framing them as unhygienic, a threat to public health and harbingers of anti-social behaviour. It's a dog whistle because it doesn't stop record store day campers. At the time of this writing, no one has been fined under the PSPO but it's there, setting a tone, sending a signal about who is welcome, while not tackling any of the causes of homelessness. There are currently tents pitched by the side of the town hall, in view of decision makers and budget holders. I spoke to Anna Minton[10] whose work has often centred on how public spaces in the UK have become privatised through policy initiatives such as PSPOs and BIDs (Business Improvement Districts). I asked her why public space matters and she said:

> It's the space and the place where we sort of realize ourselves and become ourselves because we're in public ... we interact with other people, all kinds of different people. ... If you are given the opportunity to interact with a multiplicity of different sorts of people you realize yourself in that environment, you'll also adapt to that environment, and you'll be able to deal much better with that [diversity].

Anna powerfully articulates why those interactions benefit us all, and contrasts this with the alternative:

Privatized spaces ... take away that adaptive sort of potential from people, they're not diverse, they're heavily controlled ... there's lots of security everywhere. People don't actually interact with each other and adapt to each other's needs ... they unconsciously adapt to a much more authoritarian sort of environment.

Anna's work points to the often unrecognised power of public space, how sharing parks and plazas helps us navigate through cultural differences and find our place in the city. She goes on to elaborate more about the civic values embodied within them.

In democratic public spaces we can protest, and we can do all sorts of things that we're not allowed to do in privatized spaces. So I suppose, at the heart of it ... public space seems to me to be a reflection of the public forum ... a democratic arena ... Private space isn't that. Your house isn't democratic, It's up to you to set the rules. In public we all interact together within the laws of our society. But privatized spaces change that, and they're not democratic.

I agree with Anna, and extend her vision to include streets as a place that also falls under this category. For much of the year any kind of larks on Piccadilly Garden are curtailed by maintenance or activities such as the Christmas markets which see more de facto privatisation. In May 2022, Manchester Mill estimated most of the gardens were closed to the public for 97 per cent of the year.[11] Even local authority-owned pavements are up for the highest bidder. One of my wanders last year was curtailed by a Chanel fashion show which closed part of the Northern Quarter causing widespread disruption. They transformed Turner Street into a catwalk complete with a roof and copious security. My field notes read:

I despair at how easily the rich buy our streets and close them down, but it has ever been thus in Manchester. This time it was more ostentatious and elite so of course the extravagance means glib headlines about inequality and austerity write themselves. Or rather didn't, I saw much less critique than I assumed there may be. Perhaps I'm being a killjoy again. Many local businesses felt they benefited or were adequately compensated for the disruption, but the principle really bothered me. Fashion has always been as much about money as style, ridiculous, fun, cruel and enthralling. This city has always been a catwalk and for good and bad the fabric of fashion is literally woven into our streets. The fetishisation of edginess, now only echoes remain, is hilarious, tragic and unsurprising. Just something else to talk about when discussing recuperation and commodification. The simulacrum of the NQ [Northern Quarter] has often felt just that: more photography studio than owt else, and yes, yes, I like clothes and take pictures so I am complicit in this too. There's a bitter irony that the weavers cottages on Turner Street are at the heart of the stage set despite rows about their heritage, demolition, character and managed decline.[12] Still, I do love an arcade and it's almost a shame the roof will vanish as quickly as the thrill of a new frock.

* * *

When it's not commandeered by businesses, the Northern Quarter is one my favourite parts of the city. It's where The Basement was located and The LRM was formed. Many of the buildings are the characteristic Manchester red brick and there are plenty of ginnels. There's a lot of enthralling street art, the line often blurring between advertising and graffiti. For example, I always admire 'The Giant Bluetit of Recuperation',[13] despite its commercial origin. When I first arrived this was definitely the cool part of town, but there are less artists based here now and many of the 'indie' bars are owned by the same people and share a similar aesthetic. There are still some fabulous, long established 'rice and three' curry cafes. These institutions were set up as canteens to serve Pakistani workers cheap, delicious and nourishing food. Several survive today. When visitors ask what the local delicacy is, or what our traditional foodstuffs are, I direct folk to them and invite myself for lunch if I can. Dahl, chickpeas and spinach potato, with rice and, if I'm really hungry, a chapatti for me please. When I walked with women for my research, many agreed this was a neighbourhood they enjoyed too, but there is a definite feeling it is on the cusp of cataclysmic change that could destroy the distinctive character of the Northern Quarter.

Walking along Oldham Street, from Ancoats towards Piccadilly Gardens, there are many shifting ambiences: charity offices, a chain Caribbean restaurant with a Frank Sidebottom stencil outside, vintage clothes shops, the tell-tale key boxes of Airbnbs, Piccadilly Records, Vinyl Exchange, Oxfam Books, Afflecks alternative culture emporium (previously Afflecks Palace), fashionable new bars, which I am relieved not to have kept pace with, a kaleidoscope of takeaways, cafes and ramen joints, centuries-old pubs. The Castle opened in 1776. In 1999, it became my first regular watering hole in Manchester, local not to my home but my friends' workplaces. Table football, a great jukebox and a swamp in the ladies. Umbrellas in there too in case the roof leaked again. A mosaic behind the bar commemorating one landlady, Kath Smethurst, her ashes mixed in with the grouting. A sympathetic 2009 renovation spruced The Castle up, revealed the original ceilings, fixed the leaks, added a stage, unveiling the music hall in 2010. Positive changes and you can still get a decent pint.

At number 26 is Night and Day, opened as a venue in 1991 when the area was desolate, more recently the site of arguments about noise abatement. The paradox of people moving into the city, attracted by the nightlife, then complaining it's too loud. On the other side of the road is Sachas Hotel, owned by Britannia, notorious for getting terrible online reviews from tourists. They used to own London Road Fire Station, until a community campaign forced a compulsory purchase order.[14] I've never stayed at Sachas but I have been to events. The last was the fabulous *Multitudes* Zine Fest for Black, Disabled and Chronically Ill writers. There was a DIY zine kit made by Jackie Hagan, an amazing poet and writer. She said, 'The fight for sexual equality is not between men and women, it's between people and dickheads' and her play *Some People Have Too Many Legs* was the first time I heard someone else articulate the ambivalence I have about the Paralympics.[15]

Instead of going straight on to Piccadilly Gardens, I turn into Back Piccadilly. It's probably too wide to be a ginnel, but there's another clear change in atmosphere. These are back doors, rubbish bins and cigarette butts, not neon lights and window displays. It's grimly familiar, reminds me of growing up in a tourist town, the hotels with a posh door on the front for visitors, all the rubbish stacked up at the back by the fire escape where staff would gather for a smoke and a gossip. The archaeology of Back Piccadilly is not just underfoot, it's on the walls. One door still holds a sign for the Woolworths Sub-station. The shop closed after a fire in 1979 which killed ten people and changed the law on flammable furniture fabrics. Later, Elizabeth Price won The Turner Prize with a film *The Woolworths Choir of 1979*. I went to see it at the Whitworth Gallery where people were tutting that it shouldn't be allowed.

Further down there's another pub, conveniently hidden from the main drag. There's been one on this spot since the 1870s. In 1969, it became Mother Macs. By the 2000s, it was full of Toby Jugs and knick-knacks, a good hiding place from crowds unless it was match day. Gathered there after a

First Sunday we saw the barman gifted a plated roast dinner wrapped in clingfilm and older ladies talked to me about handbags and why you should always carry a safety pin. There were yellowing newspaper cuttings on the wall, looking carefully made me squirm. I had assumed gruesome stories about the place were an urban legend but it was in living memory. The pub changed hands, was renovated, and in 2016 a large plaque was installed outside. It commemorated that tragedy: in 1976, the then manager killed his wife Maureen Bradbury, daughter Alison and stepsons James and Andrew. The plaque didn't include the name of the cleaner, Ann Hennegan, whom he also murdered. A terrible reminder of domestic violence. I'm acutely aware as I fight for the right for women to be safe on the streets too often home is no safe haven.

The pub has evolved again. It's now called The Rat and Pigeon after its most prolific neighbours. I like the new colours and recently enjoyed an afternoon with artist Jane Samuels there. We first met years ago in The Basement, where we ate tofu ice cream sandwiches, discussed our mutual fascination for psychogeography and dressed up as zombies to protest the fur trade. Later, I asked Jane what she thought about Manchester now and she said:

Manchester feels a little bit lost, it is attached to anachronism at the moment whilst it tries to redefine itself both as a slightly faceless urban monolith and something still rooted in culture and community ... it's my home fundamentally and it's where I'm from at a base level, it's about family and my personal history but I don't feel at home there increasingly as it becomes more monolithic and faceless and the communities I love begin to fracture and disappear. Some of that is about age, and some is that commerce isn't giving us a space for those things.

Jane speaks of the dynamic and intwined identities of people and places, neither are static. The impact of temporal changes and the repercussions of economic decisions are apparent in her work with the *Abandoned Buildings Project*.[16] Jane created hauntingly beautiful and surreal images through urban exploration (urb ex). She would take a cast of friends, disguised by costumes, an array of masked characters. Pooka, horse-men hybrids, angels and giant rabbits. They explored empty properties and Jane would photograph them in situ. The images were transformed into diorama, portraits of an alternative universe. Sometimes Jane would return and leave copies of her work, ghosts and guardians, a mystery to be found by the next visitors. I joined Jane on a photoshoot at a derelict school in Derbyshire. I was astonished at how intoxicating the atmosphere was, and enthralled by what we found. I had expected an empty shell, but so many evocative and personal artefacts remained. Pottery mushrooms, files full of student records, gym equipment covered in dust but otherwise in good condition. It was eerie, and I had an uncomfortable sense that at any time a pupil would come running in to retrieve a forgotten project or the caretaker would shout at us to 'scarper, go on, sling yer 'ook'.

Jane usually reconnoitred locations alone. This scouting out often felt risky because

> those places are hidden away ... people who go there, they might be artists but they tend to be folk who've gone there to take drugs or strip a bit of wall or something and there's no one there to help me if I get in trouble. In those places I become incredibly aware of my femininity, because most often the people in those spaces are men, that element feels unsafe, a little bit feral, I don't know who they are or what they might do, and I think that's the experience of a great many women in all sorts of urban spaces.[17]

Jane is right, for many women those sensations are common, part of their everyday landscape. They don't need to become urban explorers to experience feelings of risk and unease.

* * *

Many artists with an interest in psychogeography and place are drawn to liminal places, those that are on the edge, unfixed and often a bit neglected, like Jane's abandoned buildings. Julie Campbell, aka LoneLady,[18] is a musician whose records document perfectly the Mancunian landscape they were created in, and who explains the appeal of edgelands. She told me:

> my psychogeography is informed by a kind of Romanticism; the archetypal lone wanderer, frontiersman, vagabond on the edges of society, a sense of 'ruinlust'; a predilection for buildings, structures and places in a state of ruin or abandonment. The decay of Shelley's 'Ozymandias'. Ruin sites as full of poetry and potential; their structures and meanings not fixed. I don't come at this from anything like a practical place; when semi-derelict or 'shabby' areas of Manchester get cleaned up it is a disappointment to me. My sound art installation project *Scrub Transmissions* addresses these ruminations further.

Scrub Transmissions is an occasional series where Julie mortars an MP3 device into a structure somewhere in the city or its outskirts as a 'rumination on the built environment, a discreet intervention. The MP3 continuously loops a piece of music unavailable to hear anywhere else. Entombed, the music falls silent when the battery dies, and becomes an artefact / a relic / a piece of rubbish.'[19] I first encountered a scrub transmission under a motorway flyover.

37

Julie spent many years living in a tower block near that motorway – The Mancunian Way. She most definitely lived in a flat, not an apartment, and the landscape had another layer of influence on her. She says, 'I found concrete an abundant presence and this developed into an affinity with Brutalism and concrete structures, especially WWII bunkers and sound mirrors. No longer in use militarily, what now is their purpose? They take on a mystery and grandeur.' The past bleeds into the future, and it's lingering presence fires Julie's imagination, just as it does for Jane's *Abandoned Buildings Project*. The environment is an essential constituent of their work.

LoneLady's first album *Nerve Up* was recorded

> on the edges of the city centre, in a small mill crumbling into the canal. Back then it was much more of a bleak area. Industrial decay and ghosts much in evidence. I actually built a breezeblock room inside a semi-ruined shell of a space in the mill. Now it seems a crazy endeavour, but at the time I felt I needed to be in the right landscape, and in those derelict, 'satanic' outskirts I felt I belonged.

On YouTube you can find a Christmas Greeting Julie recorded in the ruins of a building she had become really attached to, the former base of Miles Platting Development Trust. Brunswick Mill is visible in the background as she explains why this is 'the right place' for her to be.[20]

The landscape continued to influence Julie, and she talks about how walking within it became part of her creative process:

> A key memory from that time of writing the record is of walking daily along the canal to the outer edges of the city, which started as a way to clear my head but became a kind of ritual in which the mundane treasures started to

reveal themselves to me and become, somehow integral to the creative process. These industrial ruin motifs are reaffirmed in the creative identity of LoneLady – looping razor wire, mesh fences, red-brick buildings, decaying buildings, concrete, corrugated iron, graffiti, mills, canals and so on. I found a language and a landscape to belong in; visual, musical, poetic, unreal, real.

That visual identity chimes with what I, and many others, cherish about Manchester although sometimes it can be hard to explain why it resonates so deeply. Opposite Shudehill tram stop, on the edge of a busy road, is a small scrap of wasteland. It recurs often in walking interviews, and on First Sundays, as a place people love. For some, it's a bit of nature and green space, for others it's a wild defiance or a way to mark the seasons. 'Kate' told me:

> I really, really like it and I'm not quite sure why. I think it might just be because it's managed to hold on and survive, although of course it hasn't survived, it only exists because a building was bought down in order to drive through this thoroughfare, build this tramstop ... I like to come and look at it and see how it's getting on.

'Kate' taught me the names of some of the plants there, said she loved this biodiverse scrap of nature, and marked the passing seasons by its changes.

I think undeveloped or unprescribed sites like this matter as spaces of potential and imagination, a reminder the city is never finished and never quite ours. The challenge can be to find an anchor, a way to understand the constant evolution of the city and how this might impact our own identity, our sense of belonging there. Walking, and rewalking, can help us to do this, immersing us in the environment, so we notice not just what changes but what remains constant. We can become grounded, and interrogate ambivalence. 'Cheryl' reflected on this, and shared her confusions around heritage, architecture and processes of regeneration in Manchester.

I don't understand why we are so attached to the Manchester grottiness really, because it is grotty and is about poverty and deprivation and what was hard for the people. But somehow, wiping that out wipes out the history of it as well and that's not fair to just rewrite the history of the city without asking the city. Like those [grotty] bits of it aren't going to be represented ... I don't think the architecture of the city actually matters ... that's the evolution of the city and we've got dirty bits and 1960s high-rises still left and we've got bits of beautiful Victorian architecture and we've got silly glass things that we thought were a good idea 10 years ago and that doesn't matter. It's the people and the scenes, the music scenes and the cultural scenes, and the sport scenes that are more valuable and worth focusing your attention on and investing in.

Ultimately 'Cheryl' articulates something important about architectural changes but concludes it is people that make the city. An actress, she also talked about how financialisation was impacting creative spaces and opportunities – something Jane and Julie also highlight. 'Sam' works as a cleaner to fund her art. Like many of the artists in this book, she juggles multiple precarious jobs alongside her creative practice. 'Sam' lives in suburban Salford, but spends a lot of time in Manchester city centre because 'I really am into cities as rational functional processes but also that kind of spiritual space where you can actually plant something and maybe watch it grow. There is always that opportunity for creation which I really love, I really love that about the city.'

However, 'Sam' is acutely aware there are conflicts and thinks:

> there's going to be more tussle ... [problems for] creative people who want to use Manchester as a kind of a creative space ... I'd like to see more free spaces. Maybe even more vacuums if that's possible, just like a space to let something happen, to sort of see what happens with the space rather than making it into a new hipster commodity bar ... [we need] the under layer that so often gets forgotten and then when it's not there anymore everybody's like where did all the creative people go? Well they got priced out!

The vacuum 'Sam' talks about are spaces where ideas can emerge and working-class creatives, like herself, can afford to experiment. She knows those opportunities are vanishing. 'Patti' grew up in Manchester and shares similar concerns. She also talks about the pressure on creative industries, always seeking something new to commodify and profit from. Her insight captures how she feels at this critical moment:

> Everybody's chasing the hip, the happening, being part of this idea of Manchester as another New York or another Berlin or whatever and that [pushes out] working-class places, as well as the rent that you have to pay to occupy a space now. It's gentrification which is social cleansing isn't it? ... it's a loss of richness I think ... because you can't have grassroots scenes that grow by themselves and grow organically through natural networks and communities without them immediately being hyped up as the next big thing.

The loss of richness, and the exclusion of anyone without sufficient economic resources, has devasting consequences for individuals and our culture as a whole. Inequality continues to grow across Greater Manchester, standing in the centre of Piccadilly it's easy to feel powerless, trapped in a cliché as skyscrapers soar up and the number of street homeless rises exponentially. This is why policies like a PSPO feel cowardly, they dehumanise and push away homeless people. The very least we can do is acknowledge their existence. Lives should not be collateral damage. Greater Manchester Housing Action is one of many groups which actively campaign for a fairer city. Amongst their many tactics have been walking tours of areas that have been most impacted, demonstrating the lived reality behind their research reports.[21]

Women will continue to find ways to make their art despite economic pressures. 'Rennie' is a musician who funds her art by working in catering. Pondering on gentrification and her continued love of Manchester despite everything, she rallies against facadism and developers who operationalise heritage tropes while destroying the actual material heritage. She wonders:

what are we looking for, why do we come here? It's invisible really, it's about what's in your imagination as well, and the whole point of a city centre is not geared around imagination is it? It is geared around buying stuff [but that's] not enough ... it is just so hypocritical isn't it, the things that developers pick up on, bits of history, bits of character, bits of architecture here and there and then they take that surface idea and use it to sell their shit apartments whilst erasing all the bits of character that were what made it in the first place so they use it and then they don't value it.

Several women spoke to me of their ongoing visceral and emotional connection to the city's histories. This was most vividly articulated by 'Kathleen', a teaching assistant who admired red brick grandeur but was aware that our riches are intertwined with capitalist exploitation: 'When I walk through the city centre and I see the really big buildings which are similar to this [on Whitworth Street] I get a ripple inside and I think about the slave trade and the funding that the slave trade bought to the Victorians. The phrase that I use is there is blood oozing out between the stones.' With this powerful account, 'Kathleen' has tuned into the largely unacknowledged horrors of imperialism and colonialism that made Manchester prosperous. If we were truly to embrace our past it would uncover not just the horrific conditions for workers in the mills and factories but also force us to confront the fact that our city profited from slavery and we have not yet fully reckoned with the damage caused.[22]

* * *

I have been trying to find a word that conveys the complex nostalgia many women discussed. They felt an acute sense of loss linked to gentrification and rapid change in the urban environment. However, they also understood that that there was no golden age; they didn't want to literally go backwards in time. All of them also talked about the benefits change had bought to Manchester and improved their experiences in some ways. This included recognition that a busier, more populated city centre felt safer and provided many opportunities for entertainment, leisure and pleasure. They appreciated this but still felt a yearning for something communal being lost. They also felt uncomfortable about inequality and displacement caused by gentrification and wanted to make sure the city did not lose its 'soul'. They saw an erasure of working-class narratives as a betrayal. For many, those memories being occluded were their own family histories. Several felt their own class position complicated by being first generation graduates or juggling multiple jobs to subsidise creative practices which were not 'proper work'. They all recognised an irreparable harm being done to the city.

Nostalgia has come to mean a powerful yearning for times past, a bittersweet pleasure or a heart-wrenching pain. It is experienced on an individual basis. The origins of the word combine the Greek words *nostos*, which translates as 'homecoming', and *álgos*, which is pain or ache. Literally, it means homesickness and was initially a seventeenth-century medical diagnosis given to soldiers who were stationed abroad. Over time, it acquired romantic associations and has come now to imply a belief that the past was better. This is why it is not the correct term for the feelings women shared with me. Regardless of their age or class they did not simply look back to an imaginary Albion, they expressed a much more ambivalent yearning while celebrating the pleasures of their present. The women I spoke to expressed fear for the future of Manchester and an active sense the city was losing something special.

The closest word I found was Solastalgia.[23] This neologism combines *olacium* (comfort or solace) with *álgos*. Glenn Albrecht and colleagues coined this term to describe the distress felt by victims of the climate crisis and other environmental disasters such as forest fires or the damage caused to their landscape by actions such as mining or fracking. The feeling of harm caused by manmade actions does resonate, and the women I spoke to felt a lack of control or agency about changes to their landscape. However, Albrecht is largely concerned with natural landscapes rather than the urban and his work does not allow for ambivalence; the changes he documents are uncompromisingly brutal.

I played around with words and thought about the idea of 'mancholia', combining the feeling of sadness commonly called melancholia with Manchester. Melancholia as a mood or affect is gloomy but also has some romantic associations. I felt this suited the colloquial 'rainy city' very well; fitting with the seductive miserabilism of The Smiths and Joy Division while anchoring unease to a specific geographical location. I also enjoyed knowing that the name Manchester is widely acknowledged as deriving from Mamucium meaning breast-shaped hill, thus foregrounding a feminine influence which has been suppressed in the mainstream contemporary associations with the city. However, on reflection this is too tied to a specific place, and we don't need another word beginning with 'Man'. I strongly suspect the sorts of feeling I want to convey are found in many western cities undergoing processes of gentrification – indeed, it may actually be a global effect, although it would be inappropriate to extrapolate wider from my interviews. These feelings are both particular in that they are clearly related to one's sense of belonging, but they are also unexceptional being a widespread and dispersed result of global processes. This means the actual physical location is mutable.

Eventually, I have settled on the word 'hipcholia' echoing melancholia but also invoking the figure of the hipster. He was referred to multiple times – like the flâneur it's a he and an archetype, not a real person. The hipster represents various facets of gentrification. For some, he is as an appealing consumer choice, with an emphasis on artisanal, neo-traditional craftsmanship. For others, he is a figure of mockery or a scapegoat. He also embodies a privileged arrogance, and a troubling social cleansing with roots that are far deeper than contemporary fashion. Etymologically, 'hip' means 'in the know', a usage recorded since the 1920s. This fits the knowing complicity in many of the processes that are causing distress amongst the women I spoke to. It also lends itself to an affective sense of both repulsion and attraction comparable to what Sharon Zukin calls 'domestication by cappuccino'.[24] Money and consumption lies at the heart of the neoliberal city and no matter how hard we may try, we remain entangled within it. I

45

feel 'hip' also emphasises that these changes are human made, not natural or inevitable disasters and they are not experienced equally or with similar intensity by everyone. It highlights the dilemma about what constitutes authenticity and value within the contemporary city. I also appreciate 'hip' has a double meaning as a body part. Hips are engaged while walking and indeed swaying hips are characteristic of the feminine gait – one of the key reasons Judith Butler says: 'a walk can be a dangerous thing'.[25]

Predominantly I use hipcholia to describe a search for the soul of a city which has no essential genius loci. It is another consequence of the spectacle which makes authenticity an enigmatic presence, equally desirable and elusive. Hipcholia is a yearning for something that never quite was, and futures lost before they were more than a shimmer in the shadows. Hipcholia is a dynamic, dialectical, emotional force that paradoxically embraces both attraction and repulsion. It's a discomfort in the relative luxury you enjoy, knowing your pleasure is complicit in historical and contemporary trauma.

Hipcholia is an ambivalent nostalgia that everyone I spoke to tried to describe. It is the search for an impossible authenticity; an essential soul that cannot ever, and has not ever, existed beneath the pavements of the city. The Situationists claimed *sur le pavés, la plage* (under the pavement the beach). It's an appealing spatial metaphor, dismantle the material realm of capitalist productivity, rip up the cobbles, to reveal golden sands. A hidden world of leisure and pleasure waiting for us (we can worry about the details, like who does the cooking and cleaning, later). However the reality is beneath these pavements are the ruins of older human constructions and deeper still nothing but rock. Tectonic plates in constant motion, imperceptible to us as we all wend our own interweaving and occasionally enchanted paths across that rock, regardless of gender, facial hair or income.

Hipcholia is what I feel as my feet trace the heart I have drawn on a map. An arbitrary line, crossing over and beyond any sensible route across Manchester. I may feel motion sickness at the pace of change, and rage at the direction of travel, but my journey, our journey, won't stop. We are all on a quixotic quest, walking our cities into being, layering our dreams, and our footprints, onto the past and into the future. I hope we can make it better.

2
Ebbw Vale

Think of someone you miss, put their photograph in your pocket and walk with them; take them somewhere you haven't been before and imagine guiding them round. What would they notice that you don't?

I tried to climb my family tree. I think it might be an oak, the one at the bottom of the garden of a South London house. We left there when I was five, I was anguished because my invisible friends, a family of mice with distinct names and personalities, made their home in its roots. I was told we could not take the tree with us and I panicked. It was OK though; my mice all hitched a lift in the laundry basket and came to Eastbourne with us, although once there, they soon dissipated.

I needed to trace my own history because I wanted to find my mother. I knew I could never actually do so, she died when I was 14, but I wanted a shade, a whisper, a story that made sense to me. When she died (not when we lost her or she went to sleep or found peace), when she died, a protective silence fell. The curtain to another room mentioned at her funeral was heavy and dense; no traces came through and she was truly gone. We didn't talk about her at home because we didn't, couldn't, shouldn't. Stoicism, tea and always doing the washing up before bedtime would get us through. If I mentioned my dead mother outside the home there was an awkward embarrassment or extravagant pity so I learnt to stay quiet about her and pretend it, she, didn't matter.

The space she left shifted in size and intensity and pain. Still does sometimes. Grief is not a straight line. What can I really remember? Images seem fixed suspiciously close to the few photographs I have but some things I know to be true. She loved Elvis, I mean really loved Elvis, and reading, and cups of tea, and conversation. She always seemed to want to talk. And I have no doubt that she cared about us. My beloved nan and father shared their memories too when I asked years later: she hadn't been forgotten, and I was glad, but it wasn't enough for me. I'm not criticising the adults who made those choices to stay silent. They did their best while wrestling their own grief and the new reality of a single parent family. It's undoubtedly a very good thing the cultural shift since the 1980s has been so huge. I'm glad to see ongoing public conversations about bereavement, mental health and dying well. There should be no stigma, no shame and no fear. We need to make sure that it's never just awareness raising and there are properly supported services too.

My life grew around her absence. A sensation, something in the corner of my eye that I wanted to come into focus. I don't know why she reappeared in my dreams a few years ago, but she did. I floundered, needed more but didn't know what to do, where to go to find traces. Somehow my ache for understanding led, as it so often does for me, to a walk. The landscape remembers what we do not. Our lives are written in the land, trails made by those who went before. If I could just stand in her footsteps it would all become clear. I recalled she took pride in her Welsh roots. I cringed when she made us wear leeks on St David's Day. I didn't pay enough attention to the linguistic scraps she taught me. When I learnt about the brutal repression of the Welsh language she wasn't there to ask about it but I felt shame on so many levels. I didn't pay enough attention to lots of things she said. I was a child. Still,

I knew where I had to go if I wanted to immerse myself in places that belonged to, shaped, her.

Ebbw Vale. I checked into a hotel on the outskirts, the purple chain one, cheap and clean and friendly and reliable in its predictability. I was immediately confused at the receptionist welcoming me back. She was convinced we had met before but it couldn't be so. I was a child staying on an auntie's farm last time I was in the valleys. I'm not completely sure how we were related: all adults, blood family, friends and neighbours were aunty and uncle to me then. However, I was fairly certain my maternal grandmother and her kin were all local (she died well before I was even conceived). The green outside my window and on the journey over was deep, dark and inviting. On the hill above is Y Domen Fawr, a Bronze Age cairn. People have lived and died here for millennia and tributes shape the landscape. Maybe it was a trick of the light but all this felt familiar, comforting. I was so close to Bannau Brycheiniog but I had to resist its embrace and instead, after check-in, went straight out again to head into town.

I was a tourist disguised as a pilgrim. There was a sparrowhawk sentinel on a fence post staring through me as the train pulled into Ebbw itself. The station reopened to passengers in 2008, having been closed for 46 years. From the platform I walked straight across the identikit plaza and campus and over to the cableway for a ride up the hill. Gleefully, I told a bemused security guard how thrilled I was to see it working as the internet wasn't clear about its operational capabilities. 'The internet lies', he warned me. My enthusiasm tickled him as I was ushered into the cabin and warned there really isn't anything to see here. I went up and as I got out felt a blast of potential, cold and refreshing. I wasn't sure where to go. I was truly dériving, following my instincts and relishing the newness. The road sign said The Walk and the portent was clear. I crossed the road by a petrol station, walking alongside a disappointingly unremarkable hedge. This, I was sure, would lead to an epiphany – but it did not. It took me to a McDonald's.

I walked past, onto the high street. A familiar mix of cafes, pound stores and a surprising number of gift shops. It's easy to buy a birthday card or bath bomb here. I sat down on a bench near the steel dragon to get my bearings. An older man came and sat next to me, started chatting because he hadn't seen me round here before. His opening line was to be grateful for benches and then he wondered, 'what bought me to this shit hole?' Perhaps I imagined the note of wariness, and how it dissolved when I revealed familial ties. He was not impressed with my job either 'because too much education gives them ideas and there is nothing here for them, all the young ones go away'. I felt echoes of warnings from neighbours in Eastbourne who told me no good would come of going to university, it's a waste of time for the likes of us. He then told me about the steel works, where he, his father and brother and more, made a living, a community, a life, and the terrible scar left when it closed. Multiple generations whose sense of belonging were connected to an industry now gone. His sadness at the lack of options for his own children was palpable, and it didn't seem right for me to argue about the value of education.

There's a museum there now, in the town hall, filled with relics from the steel works. The sheer scale of industry in Ebbw Vale was astonishing. I look at a map of the town today and superimpose what was there. Awesome; these valleys were as transformed by, and integral to, the Industrial Revolution as the Manchester mills. The museum staff and volunteers were friendly and welcoming, eager to help me. I felt embarrassed at the scant information I had, how logistically unprepared I was to find ancestors with such a common surname. Instead, I listened to their stories. There were pantomimes and social clubs and terrible industrial accidents. There was comradeship, yes, but also the hardest of hard labour, pain and grief etched onto, grafted into, the landscape.

Across the valley there were similar communities based around coal mining. I remembered mum talking about the Aberfan disaster. In 1966, a colliery spoil tip collapsed on the side of a mountain, engulfing a junior school and houses in the village below; 116 children and 28 adults were killed. For some bizarre reason, we studied it at primary school in Eastbourne, and I came home scared but excited to show her the poem I wrote. When I did so she wept, the first time I can remember seeing an adult cry, and she told me about her memories of the tragedy. Years later, I learnt that Aberfan was more than just a tragedy. It was industrial negligence and not an act of God. The grief could only have been compounded by the insensitivity of the press and the Charity Commission. Funds that had been raised could not be released to support the children unless they were physically injured, mental health was not relevant. Bereaved parents had to be assessed 'to ascertain whether the parents had been close to their children' before they were given aid. Beyond heartless.

The next day, I met a friend of a friend of a colleague at EVI, The Ebbw Vale Institute. It's the site of the oldest institute in Wales, their website dates its origin to '1849 when a group of local workmen and farmers met in the vestry of Old Penuel Chapel with the aim of forming a society for "mutual improvement"'.[1] Today, EVI is a thriving community hub with a range of services, including a community pantry and repair cafe. It's the kind of place I instantly felt comfortable in after years of visiting their kin across Greater Manchester. In 2019 when I sat having tea, I was introduced to folk who helped me decipher place names on old records, gossiped about local politics and told me where to find the best micro-breweries in the valleys. Next time! I left determined to return soon. I enjoyed my week immensely and felt overwhelmed by the generosity of strangers, but no, of course I did not find myself or my mother. I left with a lot of reading and new cake decorating tools to make sugar craft icing roses. I first used them to decorate a rich fruit Madeira sponge celebrating a 50th anniversary. Hours spent perfecting petals to show love to my chosen family as well as those with whom I share blood.

* * *

I still wanted to find my mother. Ridiculous, I know, to think I could because the brutal truth is she has been dead for over 30 years. But the silence still bothered me and I wondered what to do next. I resolved to try harder online. I found her birth certificate and Googled the address. Frustratingly, it was only around a 20-minute walk from where I had been just the week before. The map also shows a place called Chartists Cave, a more arduous trek up above the town. This had been a weapons store and muster point for local Chartists before they marched to Newport for the Uprising. Their demands were the same of those at Peterloo. Radical histories entwined, except I have no idea if any of my ancestors were involved in Wales and, to the best of my knowledge, my mother never visited Manchester. I signed up to a family history and genealogy site, because I do know she had a brother and I have cousins somewhere. I got no further than the grandparents I was already aware of. I ignored suggestions to sign up for a DNA matching service. I didn't, don't want to give my DNA to a corporation.

Instead, I hatched an absurd plan to walk from Ebbw Vale, to Carshalton, Beckenham, West Wickham and Eastbourne, following her life's trajectory, because suddenly a bigger walk seemed the most empathetic option. I would think about the logistics later, but the first step was sharing the idea. At the Fourth World Congress of Psychogeography in September

2019 I performed with Marky Daffodil for the first time in over 15 years. (We had been friends since Leeds, and I occasionally played with his band The Concrete Daffodils. My abiding memory is hiding in a KFC in Scunthorpe after a gig because we challenged some racists who then went looking for a fight with us.). The 'accessible' venue had no ramp or steps up to the stage and Phil Smith had to help me clamber on, but I wasn't going to let the opportunity go. One of our songs was about hereditary bonds and the audience chanted mum's name with me. While I kept searching for mum she wouldn't be forgotten, and maybe, just maybe, the invocation might work. My love was sent out into the world.

Then the pandemic hit and I felt strongly that now would not be a great time to foist myself on strangers or disrupt anybody's family dynamics. I concluded bleakly that they had never looked for me so maybe it was for the best. At some point, I realised it was less about her and more about the idea of her and making sense of the stories I had been told. I didn't forget of course but focused on other priorities. I locked down in a flat with my partner and a friend, too small for three people home all day but we were very lucky. I wondered if symptoms I felt were caused by Covid, menopause, my chronic health issues, a common cold or the side effects of late capitalism. I switched to teaching online, and like so many of us made new connections. I joined Micro-Climates, an online writing group founded by writer and poet Ceri Morgan.[2] The name was inspired by Guy Debord's work which views micro-climates as places of intense focused energy in urban places. It also relates to the shrinking of social life under restrictions where many became more isolated and in a heightened emotional state of pandemic living. Prior to the pandemic Ceri had been working with a dancer, Anna Macdonald, and a group of people with chronic illness, on *Circling*.[3] They had been exploring movement, walking and pain. Many of the group were self-isolating so Micro-Climates was a chance to stay together and share writing in progress. Ceri generously opened the invitation out to others. I love this script by Anna Macdonald from *Circling (again)*:

Score 1.
Choose a point in the room to arrive at
Spend ten minutes not getting there

Other walking artists experimented with ways to adapt to restrictions too. Louise Ann Wilson had already been exploring how to walk with people who can no longer physically make a fondly remembered journey. *Women's Walks to Remember: 'With Memory I Was There'* (2018–19)[4] took place in the Lake District and was inspired by Dorothy Wordsworth who wrote 'No prisoner in this lonely room … No need of motion, or of strength, Or even the breathing air: – I thought of Nature's loveliest scenes; And with Memory I was there.'[5] Louise developed a methodology where she invited people to create memory maps of landscapes they loved. They could include anything they wished – drawing, symbols, words, sound evocations, photographs, whatever they liked. Louise would then walk the route and share her surrogate experiences with the mapper. During Covid-19 restrictions this practice evolved because, as Louise says, memory mapping offers a 'creative, therapeutic and communicative tool'.[6] Participants don't have to physically make a journey but can instead enter landscapes through their imagination. Memory maps could be made by individuals or dispersed groups separated by the pandemic. Louise invited participants to share their maps, a short text and three related photographs with her. These were collected into a gallery which remains an exquisitely poignant momento of an extraordinary time.

The pandemic wasn't, still isn't, over, but, I was able to return to Ebbw Vale in 2022. I decided first to take solace and inspiration from the owl sanctuary, the opposite direction to the town. The next day I planned to go to the Rassau and see the house mum lived in. I wasn't sure if I would knock on the door or just stand and stare. It was a gorgeous sunny day and I felt once again freedom fizzing in my blood. I walked through some houses on my way to Festival Park. The hills were calling but, again, I never reached them. Scuppered by gravity. I tripped over a broken slab and found myself sprawled over the pavement, an instant that changed everything. I tentatively realised nothing physical was broken, no harm done except to my dignity and inelegantly pulled myself up. No one saw me but still I felt embarrassed. There was a deep ache in my leg but I resolved to carry on. A lifetime of falls means I instinctively know how to land as safely as possible. I was limping but OK and my quest would not be deterred.

I entered the park, a previous Festival Garden site like the one I know in Liverpool. Near the entrance, in a section of lawn surrounded by hedges are a series of weathered statues, eerie and decayed: a boxing glove, a baby and a skull amongst them. A graveyard of dreams. The park was deserted, and for a moment I felt totally alone, and as if I had discovered something for the first time. I continued up the hill past the fishing lake and sat on a bench with stunning views to admire the greens; Welsh green is so verdant. Shamrock, emerald, moss, chartreuse. I let myself dream a little. Then I realised I had forgotten my water bottle, dropped when I fell. It was hot and I was dizzy. The cafe was closed, and the aching was getting worse. I cursed my lack of preparation. Townie! I realised the owls would have to wait. I took myself back down to the hotel. I swerved into the pub and ordered a large glass of water followed by a pint of Guinness and an ice cream. Sorry for myself, I needed comfort, but this cavern with a kids soft play area attached was the only place I knew open.

I sat outside to enjoy the late afternoon sun, recalibrate, and rue my idiocy. Then, I tried to stand up. I could not. My leg had locked and it was agony to put weight on it. I leant heavily against the wall, and must have looked distressed as a biker asked me if I was OK; for reasons I don't understand, I said yes and declined help. The few metres across gravel felt an unsurmountable trek. I hobbled, shuffled, propelled myself across to the hotel reception. Struggling not to cry, I asked on the off chance, casually as I could, had anyone left a stick or a large umbrella behind? They hadn't. And no, they weren't allowed to share painkillers. Leaning against the wall I inched to my room. Navigating became a nightmare: the bed was too high and the bathroom too far away, but I made it. I knew I needed to eat, but frustratingly couldn't get food from the restaurant because they had no remote payment facility and I really wasn't sure I could make it across the car park and back again. I suddenly appreciated delivery drivers and over-inflated convenience store products.

This fall was caused by a damaged paving slab and gravity. However, falling was part of my diagnosis as a child. My mother knew something was wrong because I fell so often. I grew up knowing instinctively why a helping hand matters, that try as I might I could not always do everything I wanted to do by myself. Perhaps falling is why I went on to community-based work, because to fall is to understand that we are all co-dependent and interconnected and need support sometimes. This is integral to the experience of many disabled people: our support networks and sense of co-operation and connection. In an audio-collage of LGBTQIA+ people who experience energy limiting conditions, Mish Green talks about the *Cripothecary*.[7] This describes the ways we cope and care, and the folly of well-intentioned but frustratingly useless advice from people who think they are being helpful. I hate the word resilience because it has been corrupted by politicians to the point of meaninglessness: our policies will hurt you, but here are some ineffectual tools to pretend we care. But actually, disabled people are by necessity resilient. We have to adapt to change and the new obstacles always appearing. This can be incredibly hard, but we carry on. Artist Jane Samuels[8] told me about how she is adapting to chronic illness: 'For the longest time I've really defined myself as a hiker and a walker and a lot of that has been about pushing myself to my limit … I'm having to redefine both my practice and something about who I am and what I do.' Jane talked about having to rethink plans to conduct walking interviews with people on the Fells because she has begun to fall more frequently. She acutely feels multiple levels of responsibility to herself, to other people she may walk with, and to potential rescuers. Jane's relationship with her body has changed, and so 'I have to be sensible in ways I'd prefer not to be. I have no choice.'

My partner drove to collect me to save me taking the train journey home, and I felt the immense relief of being cared for. I haven't been back yet but Ebbw Vale remains a precious place to me. It's a bittersweet ache and a promise unfulfilled. It's a home I never had and a place that I don't know. I will be back, not to look for mum but because I want to visit the owl sanctuary.

*   *   *

The 2024 Fourth World Congress of Psychogeography was held in Canterbury. Jane Samuels and I travelled down to it together. The event had *hiraeth* as one of its themes. Their website describes the meaning of this untranslatable Welsh word as 'deep longing for a person or thing which is absent or lost; yearning; nostalgia; spec. homesickness'[9] and this chimes with me. However, I do wonder if I am appropriating a culture somewhat. Ebbw Vale has never been my home and my roots there are not strong. My mother left some time in (I think) her late teenage years and I haven't yet found any family still there. I wonder what it is about valleys that makes me feel safe. It's probably more about The South Downs than the mountains but I have a suspicion of flat landscapes. I always want to see a hill I can run to, despite the fact I can't run.

I think about the work of Elspeth Owen[10] whose work is on a much larger scale than mine but who explores friendship and connection. Her first 'Big Walk' was to the Greenham Common Peace Camp, a place pivotal for so many amazing women. Her 2005 walk *Looselink* was a sequence of hand-delivered messages – one person giving her something to pass on. She didn't know the participants or where they would send her. In *Grandmothers Footsteps,* she walked 15 counties to hand deliver messages in a similar chain between people who shared her new role as a grandparent. There is a faith and trust in strangers that shines through Elspeth Owen's work. Her score for *Hop, Skip, Jump* is

Hop, Skip, Jump

I walk to a friend (no more than a day's walk from me) and stay over, the two of us walk together (no more than a day's walk) to a third person, (unknown to me), we stay over and then the three of us walk (no more than a day's walk) to my place.

It would also be possible to do a version which all happened in one day or two days, according to the time people have.[11]

There's an episode of the BBC Radio Show *Open Country*[12] which talks to Elspeth Owen about her *Blue Moon* walks. To mark the unusual phenomenon of two full moons in a month Elspeth remains outside the whole time. At night, she walks to bury ceramic necklaces she has made. The interviewer seems astonished and keeps emphasising her age – Elspeth was born in 1938 and gently mocks the journalist for bringing news-papers. Elspeth sees her work as rebellious, and I find the faith and trust in people she embodies so moving. She told Dee Heddon and Cathy Turner another motivation is fear she feels walking alone in unknown places – but the bad things are imagined: 'somebody has probably done something fantastic for me, or shown me the way or taken me in, or I have heard nightingales'. The first time Owen slept on her own, outside, in just a bivouac, she awoke to the sight of a white stag: 'I think I know these things are there and I could enjoy them if I dared and so then I dare and there they are.'[13]

Rosana Cade's *Walking: Holding* also explores issues of trust, alongside questions of identity and belonging. Their work begins with the intimacy of holding hands, and what it means to be able to do this in public. Six different people in turn take a single participant on a walk around part of their city, holding their hand while they do so. Rosanna explains the origin of the work in experiences with their first girlfriend, and how, in queer relationships 'Hand-holding always felt like a complex act – the tussle between visibility and risk, public and private intimacy, activism and fear.'[14] The roots of this walk are in the LGBTQIA+ community, and navigating an environment where homophobic and transphobic attacks remain a real threat. They say *Walking: Holding*

> gently encourages people to abandon fear and trust the hands of strangers. What emerges is often uplifting and empowering. While the awareness that it is a performance allows for this abandonment, it is also 'real life' and in a public space – anyone en route could be a potential hand-holder. I am interested in turning the stranger into a human, turning the 'other' into a human, and, yes, seeing a wider range of queer hand holding.

There are so many ways to love and care. I wonder how much blood counts, and what is in our DNA. My genetic inheritance was another reason to wonder about my mother, and developments in medical science made me more than curious. My diagnosis has fluctuated, like my pain levels and ability to balance. It's still unclear, which is one of several reasons why I don't share it, too many experiences of well-meaning Google detectives. I learnt in my mid-forties that whatever it is, it is definitely progressive and carried in my genes. That's something she gave me that I can't ignore, and perhaps that knowledge is why I went in search of her when I did. Perhaps my future could be revealed if I walked her past. My mother was not a saint or a fairytale. She was a human, and she is with me in genes and dreams and half remembered stories. She did cross-stitch, made tapestries that became cushions and table-cloths and pictures, and once she made a rug. I still have that, one of the few relics left. It is a steam train chugging across my bedroom floor.

I'm thankful for the reframing of craft by walking artists like Lizzie Philps.[15] Her *GPS embroideries* are an 'invisible mending' of the countryside. She thinks about how GPS signals criss-cross like stiches. Walking to 'embroider' the landscape takes the domestic outside to write large across the landscape. Beautiful. This is public art at so many different levels and Lizzie has facilitated workshops with many groups of women, girls, carers and mothers. I've also learnt from the walking art of Elspeth 'Billie' Penfold, an artist of Bolivian-Argentinian heritage. Her work blends walking, weaving and poetry, and she tells me 'my work is always about process ... that process goes on it ripples, it doesn't finish at the object'.[16]

When I walked with Elspeth Penfold at the Fourth World Congress of Psychogeography she gave everyone in the group a piece of rope she had woven herself. She introduced us to the walk *Lucky Dip: Diviner* by thinking about the Quechua language which she heard as a child. Their word for thread is the same as that for word, and embroidery is a complex conversation. There is no concept of a singular, isolated being in the Quechua language. This reflects Indigenous Andean cultures where, as Elspeth told me later, 'You only exist in relation to other things, you do not exist as an individual. The individual isn't there ... your existence is in the relationship.' The example she gave stayed with me: the space between fingers, where they join, has a name but individual fingers do not. We drifted around Canterbury exploring the theme of community spaces, some of the group had been involved in a local charity shop and support group. Occasionally, we stopped to read poems Elspeth printed on ribbons, we also stopped for her to greet friends, there were many serendipitous encounters. Later, I spoke to Elspeth Penfold about how her heritage speaks through and in her work.

She told me she first came to the UK as a student to study sociology and a lecturer

said to me you're an interesting example of a marginalised person, and I was only 19 and I was quite disorientated ... although I could speak English, my English was very good, that linguistic ability has denied me of my background ... I have carried that with me all the time I have been in the UK, this marginality.

Developing her walking art practice *Thread and Word* [17] has, she says,

allowed me to connect to my heritage hugely and to come to understand it better ... we all carry our culture with us but to be able to turn around and examine it as part of your practice it lends a much deeper understanding of so many things that you had in you.

At the end of our walk in Canterbury, as with all the *Thread and Word* walks, participants all tie their ropes onto a single stick. These are inspired by Quipu and Elspeth has a growing collection which tells the story of each walk. Quipu were a central and sacred part of Inca culture. The knots and textiles are a language used to document valuable information. Elspeth laughs as she says how many academics have tried to decipher the meaning, and she herself does not know the code. She says, 'I'm not all that worried in cracking it, but I am interested in how it makes people feel, how walking and knotting ropes can help you feel part of something, involved, it brings you in.' She thinks about how so many words that connect us to the land and body are about weaving, and she also tells me she does not unravel if she ever makes a mistake while crafting because 'it's not what you planned, but it's not a mistake, you incorporate it into the weaving'.

Listening to Elspeth Penfold, and sharing her walk, I learnt much about how we might walk with our ancestors, to honour our heritage, appreciate our culture, but also how we walk into being new communities. She is an active member of the Walking Artists Network and shared 'being part of WAN really bought it home to me that I am not marginal, I am a person who belongs to the earth and of this earth and that we all belong'. I hadn't actually heard of 'walking art' at the dawn of The LRM, and when I first connected with the Walking Artists Network, it made me too feel part of something bigger but I still haven't resolved all the mysteries of my more personal inheritance.

I was well cared for as a child, and always told my feet were a mechanical fault, fixable and stabilised as long as I was careful. But I was being shielded. If I look past the birthday parties and Christmas presents and Sunday roasts I can see my mother was not happy or well for many years. I didn't know the word 'agoraphobia' but I learnt it from a teacher who asked if they should visit mum at home instead of expecting her to attend a parents' evening. I still don't know if they should have asked me that, but they did and learning the word itself was revelatory. I was always a serious child and I don't recall me daring to ask any questions because relaying that offer did not go down well. I don't even think I had really noticed before how mum didn't go out much, because children only know what we know and think everyone's family is like ours until we don't. My world radiated out from our house, and she was always there, and it didn't really occur to me to think anything was wrong with that.

The story my nan told me was that mum got afraid after falling over while pushing a pram; she didn't want to hurt her children. A story of maternal love and care and sacrifice. I know now how women's health is met with disregard and disbelief by medical professionals and I can only imagine how much harder it was in the 1970s and 1980s. I believe she was pathologised and unfairly treated. Her physical pain ignored. That hurts me most because I know how much the right prescription drugs and orthotics improved my quality of life. She withdrew into the home, the domestic sphere which centuries of tradition socialised women into. A woman's place. She became the thing that, when she died, I demonised, thought I hated and feared most in the world: a housewife. That's probably the hardest sentence of all for me to write because I fully acknowledge it was internalised misogyny that instilled the notion that home-making is not worthwhile. It took me a lot of work to dismantle that false belief and truly understand the value of what she gave her family.

Mum's death was sudden, shocking. She wasn't well – well, much more unwell than normal – one Tuesday morning. I waved her off in an ambulance and went to school, not really concerned because hospital fixes you. That had always been my experience of surgery: lots of attention, kind people coming to read stories to me in bed, and a new doll to keep me company each time I had to stay on the ward. I never saw her alive again. My nan held her hand. Years later, I held nan's as she died too. My motherless life has been filled with amazing women and she was one of the finest. But that's another story. Grief, like time, is not linear, it's a monstrous spiral that can re-entangle you years after things seem settled. It's a hole that skin, and life, and memory grows round. Still, occasionally, more than half my lifetime later, an unexpected song can conjure a vision of mum and leave me gasping. She was cremated. There's no memorial but her name is in the book of remembrance. I don't know why but I've never gone to see it. She's not there. Instead, I think of her dancing in the kitchen, helping me with my homework, or sat in the car at Holywell, watching a storm. I thought a picnic meant sandwiches in the car because we never got out. I celebrate her birthday but don't commemorate her death.

I feel most connected to my mother when I cook, she loved (or I think she loved) baking for us. I make Welsh cakes like she did, but mine are vegan and so this is not a family recipe. I have no idea what she would think of who I became, or even if we would have made it through those difficult teenage years without tearing each other apart. I may appreciate *hiraeth* but I still have no time for its cousin nostalgia. It can stop us looking forward or clearly appreciating what we can do, must do, better. Nonetheless, when I rub margarine into flour and stir in mixed spices I feel part of a tradition, a lineage, and there's a pride I never quite understand because no one chooses where they are born or who they are borne of.

I do know one thing, clearly and with no regrets. This branch of the family Rose tree ends with me. I am the end of the line. I decided not to have children of my own, a decision I have never regretted despite revelling in my own role of aunty (honourable and actual). Giving birth sits in my body alongside climbing a mountain, running a marathon or being on *Top of the Pops*. Wonderful things to do, and I take joy from the joy of those who have made those things happen, but it's not for me, not in this life. And that's OK.

The irony is not lost on me that mum retreated inwards and I obsessively went out. The walking artist who couldn't walk is a paradox of my own creating but I won't stop. I still have so many questions. When I asked something tricky she would say, 'that's for me to know and you to find out' but I probably won't ever manage to do that. I wonder what she was like as a child or as a teenager. In the few photos I have she looks lively, vivacious, groovy in a beehive and mini-skirts. I wonder what of her is in me and if my deduction is right, these genes were hers too and she was failed by misogyny and men in white coats who think they know better than anyone. I wonder what really stops me sharing a DNA sample or knocking on random doors. And, perhaps most of all, I wonder how she walked.

I drifted away from my own script here, didn't follow those directions exactly, and I don't expect you to either. There wasn't an actual photograph in my pocket in Ebbw Vale as I feared a relic would get crumpled or put through the wash. I don't think I need it – the dialogue is in my head. If I do fancy a visual nudge, there's a galaxy of images on my phone. We all walk with ghosts, ancestors and descendants wherever we go, it's whether we choose to let our imaginations tune into them that determines the conversations we have. Their influence may be subtle but it's not supernatural. Wherever I walk now my mother and my nan are here, in my genes, my dreams, my wayfinding and my wonky footprints.

3

Eastbourne

*A treasure hunt. You can do this wherever you are, inside or out,
and the duration can be from three minutes to infinity. Please
make of it what you wish. Search for*
Wild Music
Futuristic Smells
Star Children
Better Times
Street Poetry
Connections

There is an assumption that walking is simple: one foot in
front of the other, easy does it, primal, instinctive. This
assumption is a fallacy that all bodies are alike and walking
comes 'naturally' to all. This wasn't true for me. I hated
walking in Eastbourne, the place where I lived for most of my
childhood and teenage years. I would get scolded for crawling
upstairs or slithering along the floor which felt more normal,
and comfortable to me, than picking up my legs. There are so
many words for walking, some more acceptable than others.
Eli Clare rallies that 'sliding, scooting, crawling, crab-walking
… gimpy walking' are not talked about or appreciated, even
among disabled activists. 'Falsehoods – *burden, clumsy, better
off dead, tragic, dangerous, not fully human, childlike, worthless* – reverberate.'[1] In my teens, I internalised ableism and
would be mortified if anyone suggested there was something
wrong with my walking. I wore gold Doc Martens boots and

stomped, strutted and stumbled along ignoring the pain. Years later, another doctor categorised how I move as a 'waddle'. That word would have crushed me then, but now it's OK, it's true. I have come to understand, and celebrate, many different kinds of wandering.

I don't actually remember my first steps, but I do recall using a walking frame which my younger brother enjoyed playing with. Later, I can still recall vividly endless processions in front of doctors at Guy's Hospital. They were apparently fascinated by my gait and they weren't alone. I must have been about nine, in a drama class, when we were all instructed to cross the floor as bears or elephants or mice or squirrels. Then, the invitation came to follow another pupil and try to imitate their movements. I became aware of a gaggle following me, moving their arms around aimlessly and swaying as they crossed the floor erratically. As various strangers have helpfully reminded me across the years, I'm a Weeble,[2] a penguin, The Penguin, an oompah-loompah, obviously drunk,[3] an in-the-way, hurry up, tut-tut-tut why do they let you out disgrace. I walk funny.

I found the mockery easier than being patronised. I knew how that felt before I could read the word. Sports Day at infant school, made to take part in a race. Parents clapping and cheering on the winners and I had barely left the start line. They cheered me too, gave me a special prize, but I knew this wasn't quite right and somehow I was a bad kind of special. The wrong kind of medal. I got very good at making excuses and missing games after that. I suspect it was easier for everyone so, by the time I was at a secondary comprehensive school, no one ever questioned that I had a period almost every Wednesday for five years.

I grew up on a street with no cars, just a pedestrian walkway. Magnolia Walk. A beautiful name for a path, conjuring a paradise rather than a suburban estate. Perhaps this freedom from traffic shaped me more than I appreciated, although it

was a freedom seldom realised. I was kept safe ind
I used to sit on my bedroom windowsill, betwe
and the curtain, legs curled tight, radio on loud, ai
world with an intense yearning. To my left the South Downs,
seemingly beautiful, wild, endless. When I realised these were
not in a 'natural' state but carefully managed and maintained
I lost a little of the magic in my world but only until I next
roamed up there. Elementals can still frolic in an agricultural
landscape. To the right a path, past neighbours, over the road,
down the hill, skirt the fields at the back of the community
centre and then on, on into thrilling adventures – the city:
lights, danger, reinvention. I longed to explore, to escape, to
do something, anything. To grow up and get out.

My walk to school was a painful slog. We lived where we
did precisely because of my legs; the school was visible from
our front window and well within my capacity to dawdle over
there. I didn't get much chance for mischief on the way home
because my mother could watch me through the living room
window and, with the all seeing eye of a parent, she would
know if there was any cause for concern. I wonder now did
this sow the seeds of my interest in the impact of surveillance? Maybe I am so committed to loitering now because I
wasn't able to back then. Cycling gave me a greater level of
freedom and speed my feet could only aspire to. Years later,
I revisited the 'mountain' we zoomed down at great peril and
realised it was just a slight incline. I dare not go back to the
local park in case the forest reveals itself as a copse. I don't
cycle any more but disability cycling advocates such as Harrie
Larrington-Spencer[4] campaign fiercely for cycling justice and
reminds us that a bike can be a mobility aid too.

I can recall two wishes I made repeatedly as a young girl.
I understand the pact you make with birthday candles, swap
the light for a secret, share it and it can never come true. I am
good with that now, glad actually I didn't get my heart's desire

as both now seem so perverse and wrong. I asked to be pretty, to have a body that attracted not repelled, and offered to swap being book clever for the chance to be 'normal'. Too many adults told me 'it's a shame, but at least God gave you a good brain' or similar. I didn't want to be stuck in books, I wanted to fit in, to go outside. I also wished for adventure, and I am not sure what I meant exactly. I don't think it was peril or high-octane action necessarily but it was for something, anything to happen. I wanted a purpose. One of the abiding emotions that rumbles when I think of life then is boredom. Like multitudes of young people in suburbia, I was never quite sure what I was waiting for but I would be ready when it arrived. Years later, I would learn the SI were 'bored in the city' too. [5] All those years of incubation were not wasted, they were an apprenticeship.

There was miles of aimless and wistful walking with friends throughout my teenage years. Slow and filled with many stops to chat. Basically, we were looking for that elusive some-thing, anything, to happen. Too young for pubs, too stifling or rowdy at home, what else is there to do when the park becomes boring? Cruising down streets, sometimes passing acquaintances and classmates heading in the opposite direc-tion. They didn't know where the action was either, but no one would ever admit this. Being a teenager is not a crime, and young people are not the problem. Adult violence simmered in the background everywhere outside my home, but I care-lessly never made a link with concerns my parents voiced. The casual slap into the gutter, the threat of the Big Kids on bikes down at the rec, the neighbour with a knife shouting at his girlfriend. It was just normal. We turned out as teenag-ers dressed up for seduction but not knowing what it meant. Shoes too big, shoulders uncovered, shivering, quivering, gasping for attention that too often came with harmful intent we were too naive to spot. There were so many petty feuds, transient friendships forged and broken in bathroom mirrors.

Comrades abandoned at bus stops after one too many ciders. Find your own way home. Maybe books and records were better companions after all.

I learnt my gender limits here in Eastbourne in so many ways. The first time I was flashed at I was shaken and upset, and reported it to the police. They shrugged their shoulders, 'an epidemic' apparently across Gildredge Park, Hampden Park, Old Town. When it happened again I was angry, and the third time utterly furious. This felt worse: I was followed, targeted, ambushed in a graveyard. I didn't bother reporting it though as I knew that would just waste everybody's time. Instead, I rang a friend and her mum made us large gin and tonics while teaching us witty retorts and ways to stay safe or at least alert. This happened in Eastbourne but could have been anywhere, any time in the UK (and beyond). Everyday sexism and harassment are so embedded as to almost be banal.[6] This is not OK.

* * *

The need to protect ourself is embedded in daily routine and women's 'safety work', as Liz Kelly[7] calls it is taken for granted. This became a theme in my thesis research because it is impossible to ignore. The confident stance, the ungendered clothing, the keys in the hand or the alarm in the pocket or the phone always ready. Will you make yourself small and try to vanish or shout as loud as you can if you think you are being followed? There is no one right way, and even if you follow all the safety advice, you can't always protect yourself. 'Rennie' is a musician who told me 'I actually run with a key my hand, between my knuckles, you'll probably be okay but ...' . Her voice trails off and I know why. Her cautiousness is based on her lived experiences as a woman, and has a profound impact on how she relates to the places she lives. 'That's a big regret

I have that I feel like [I] can't be quite as intrepid as I would like to [be]. I would love to walk alone at some strange time of 4 a.m. just to see, to get a sense of it as otherworldly at an odd time of the morning, but I would be crazy to do that.'

'Rennie's' explorations are curtailed by a fear that is justified and reinforced by stories from friends and family and media reports. 'Layla' lives in Manchester city centre and reiterates temporal as well as spatial limits 'I think everybody relaxes when they are near their home but ... I've got a map in my head of where it's okay to walk on a Sunday morning and where I wouldn't necessarily walk on a Friday night ... I'd go round the long way, I'd go home that way.' This knowledge we carry with us was summed up by 'Veronica' when she told me 'walking and loitering for men and women are completely different, I think men loiter and they can do that I think, if a women hangs around somewhere for long enough something is going to happen so I think women are constantly always on the move'.

This is expanded on by Julie Campbell (aka LoneLady from Chapter 1 'Manchester'), who says

> Of course; women are simply not free to wander as men are ... Can you even imagine it – wandering nonchalantly in the dark alone? ... There's far more danger for women, and as a result there is a high level of vigilance that has to be activated. Which makes lone walking less free, less pleasurable ... As I write it's Winter, it gets dark early and I have to modify which routes I am walking from say, the shops, a meeting or the gym to back home, and if there is a slightly emptier, lonelier stretch then the vigilance levels are higher. I check the bushes, my peripheral vision is on high alert, I don't walk past parked cars, I walk in the road. And that's every day – just ordinary walking, never mind walking as adventure, as exploration, as practice. I don't think men

grasp the extent of vigilance and strategic planning of ordinary activities women have to deploy on a daily basis. The streets, and the practice of exploration is a very different prospect for men and for women.

The weight of this knowledge does not stop women walking. They enact resistance every day, whether consciously or not. 'Nora' speaks for many when she says,

> I refuse to be scared. I really feel very strongly about that and I'm like that with my daughter, about being able to walk around and having realistic sort of feelings about what's going on around you, not being too scared about what's going on ... but I always do think about how I'm getting in and how I'm going home.

That constant vigilance and safety work becomes exhausting, reinforcing the myth of the impossible flâneuse. Many women walking artists have explored the potential of night walks to open up space and share opportunities to explore with each other. The first walkwalkwalk events Clare Qualmann organised with Gail Burton and Serena Korda were night walks in East London, creating what they called an 'archaeology of the familiar and forgotten'.[8] The group moved through the city together, exploring and chatting, sharing soup around a brazier in an underpass. Later iterations followed a similar route, adding new layers as some participants had been previously. Fly posters appeared on walls they had walked past, each print containing stories shared. As they weathered, those stories became embedded into the environment. Juliette Adair is quoted on the walkwalkwalk website sharing her experiences of that first walk: 'it was somewhere between walking to work, a dream and a party ... It almost felt like direct action. Reclaiming something ... It made me think of the three of you, individually, walking in the weeks before. Women walking alone.'

In Bournemouth, FoxStrut is a joyous event, where women and members of the LGBTQIA+ community dress as foxes and parade through the town demanding their right to be there, to take up space gloriously. Co-founder and photographer Jayne Jackson says, 'Bournemouth is actually the UK's capital of urban foxes. We see them all the time and after dark we probably see them more than women on the streets. This is about making social change and coming together and making a difference.'[9]

Zarina Dolan's *Running Shoes*[10] is an audio walk, designed to be listened to while walking, It brings together women's voices talking about what they need to do to feel safe. Dolan was based in Glasgow when she created the work. It was partly in response to the murder of Sarah Everard which bought the safety of women in public to the forefront and inspired the Reclaim These Streets movement. *Niqabi Ninja* by Sara Shaarawi[11] was also produced in response to specific events, in this case sexual assaults by mobs in Tahir Square. The audio-led journey incorporates site-specific street art into 'a graphic-novel style revenge story about one woman's transformation as she attempts to right the wrongs of the male violence she sees all around her'. It is designed to be experienced in pairs or small groups and she describes the work as 'Intimate, visceral and pulsating with a soundtrack to inspire your own resistance against gendered and state violence.' Both *Running Shoes* and *Niqabi Ninja* were produced under pandemic restrictions and also illustrate how artists adapted to different kinds of limitations.

There is a grand tradition of mass demonstrations of women's determination and resistance. We demand the right to space by taking up space together, and the power of walking together is electrifying. Reclaim the Night, SlutWalk, Pussy

Walk, Hollerback – all of these events are profoundly political, but also creative and inspiring in many ways. There is joyful solidarity in making banners and placards, meeting friends, making new ones if only for the duration. 'Enough is Enough'. 'These are our streets'. 'Whatever I wear, wherever I go, Yes means Yes and No means No'. There is power in collective walking. Change is slow and one step forward, two back – absolute agony but we will rise. I will never forget the electric jolt I felt at my first Reclaim the Night, it propels me forward even now.

This book is focused on the UK, but the safety of women walking is of course an international concern and networks of events provide support and solidarity. *Women Walking, The City, At Night* curated by artist Eléonore Ozanne[12] began in Seville, Spain, as a collective performance walk at full moon. They now take place in eleven cities in Spain and internationally. Women Walk at Midnight[13] began in Delhi, India, by Mallika Taneja, where there have now been 55 walks, along with sister events organised by chapters in Bangalore, Faridabad, Guwahati and Noida and Cape Town in South Africa. Their website articulates perfectly why this matters:

We do not seek permission to walk. We just walk.

We believe that we are as much citizens of our cities as the men who pass us by at night. The roads – long, narrow, wide and short, with their potholes, bylanes, broken pavements, well lit and not, with their trees and flyovers and the constant din of construction of ever growing cities – are also ours to occupy. We walk, when we want to, where we want to.

These are not organised protests and yet, this does not mean that the walks are not in protest.

This is a way to resist and push back on what is socially and culturally 'disallowed' to women. We take strength in our togetherness, wear what we want, often put on some music, and meander through our city.

The solidarity and comradeship of a collective walk of resistance is a powerful force, the energy is contagious and points forward to a better, more just way to walk. As The Women Walk at Midnight proclaim

We walk, because we can.
… in the hope that by walking continuously, we will finally walk into the day that each woman can walk, on the street, alone, at midnight. Everywhere.

The hope expressed in their statement resonates across continents. In each place the cultural climate, and specific experiences will differ, but the challenges women face is shared everywhere. At the heart of every action, and every artwork, is a woman. Kubra Khademi's performance *Amor* is one of the most remarkable. She wanted to highlight sexual harassment in her homeland Afghanistan. She walked the streets wearing a costume she had constructed which emphasised her breasts and buttocks; the feminine signifiers of her body. The costume served as both protector and accusation. She walked in Kabul, in her neighbourhood, 'I wanted these people to be my audience,' she said. 'I did it in Koti Sangi because it is a real place, to say what happened to me many times there, and the rest of Kabul as well.'[14]Crowds gathered, jeered and threw stones. The threats and abuse she received resulted her being forced into exile in France where she continues to make art. All these women who led acts of creative resistance offer a powerful alternative narrative. We refuse to be scared, we will empower ourselves, defy the edge of the map and walk together for change.

* * *

When I walk in Eastbourne now I carry these artists, and these acts of sisterhood, solidarity and rebellion, with me in my heart. Walking when I was a teenager may have been risky but it often felt a chore. I tried to save money so would often undertake what felt like tedious long treks to college, to shops, to visit friends so I could avoid spending cash on bus tickets. But also, somewhere, somehow, I was beginning to appreciate the wider affordances of walking. Almost every lunchtime I took the same walk for respite and inspiration. I was working in a shop selling toiletries and cosmetics. It was not terrible. My colleagues were mostly great; supportive, funny, fierce. They were kind to me in ways I was too young and obstinate to appreciate at the time. I learnt a lot from them all about what it means to be a woman in our society. Lipstick can be many things: a mask, a weapon, a shield, a performance, a comfort, an artistic statement, an expectation, an imposition, an invitation, a habit, a uniform, a treat, a defiance. I also enjoyed talking to customers (mostly); the nature of the products meant they tended to be bought as gifts or indulgences or to make the buyer feel better. But, it wasn't where I wanted to be.

Leave the shop, turn right. Go past the kitsch faux heritage store and wish you could afford lunch from the deli. Resist rummaging for charity shop treasures or second-hand record shop bargains and head down Terminus Road. I mentioned its allure to a friend in Leeds once; she assumed this was a euphemism for suicide. No. It's the way to the sea. Past pubs which open early, sell Bovril and Harvey's, then the restaurants that seemed to me then to be stuck in time. I love them for that now; charming ice cream parlours where friends' parents met and the fish and chip shops that never change. There are souvenir shops too: rock and fudge, postcards, London memorabilia and assorted plastic tat. One used to have a shelf of vibrators above the counter which baffled, embarrassed and then amused me at different points in my life. The air is salt and vinegar, the horizon blue and grey mutations.

Cross at the zebra and look at the carpet gardens. There used to be a topiary spitfire which always troubled me. Too jolly. I remember nan, who loved a shindig, recoiling from street parties commemorating the anniversary of VE Day. She told me she remembered the day the war in Europe ended well, and it was amazing, though she couldn't celebrate fully because she didn't know where granddad was and the GIs were too frisky. Like many of her generation, she never went into details but told me too many had lost too much, and too little had changed, for jingoistic bunting and buns to be appropriate now. I wear a white poppy in November. To the left is the pier which holds an important piece of my heart. When it was on fire I felt the pain although I was hundreds of miles away. My sister told me there had been racist complaints about the restored gold domes looking too 'foreign' and 'exotic', and I was proud to hear her challenge the nonsense. Pleasure piers always fascinate me, the magnificent folly and gleeful pointlessness of them. Perfect for loiterers. You can guess an Eastbournian's age by what they call the nightclub. I was a Roxy girl. This was a formative place for me, the lights and the darkness and the unwanted, uninvited groping and casual sexual assaults. A rite of passage nobody should endure. The me of the lunch break walks did not go onto the pier, no time, instead she turned right along the upper promenade. There was a definite showing off in the teenage strut, measured differently to the older ladies sat on benches and the families with kids. Go as far as the bandstand, imagine being on the stage, salute the cliffs and then quick as you can cross the road and head down Cornfield Terrace past the war memorial and wine bars, then back to work.

Retreading these steps now my feet know which way to take me. Many of the shop fronts have changed. Today, the pub that sold Bovril is a branch of Bills, Debenhams is rotting away empty, many of the restaurants have changed hands and I see lots of sourdough and fusion menus. Still so many enticing charity shops. I'm surprised how viscerally I hate the new Beacon Centre and wonder what right I have to complain. It's an identikit mall, an extension of the Arndale Centre, which swallowed up a number of independent businesses. In terms of economic strategy, it seems the wrong move at the wrong time, but it's open late and I suspect teenage me would have rather liked it. I'm still thinking about this as I reach the seafront. In my teenage routine I would go right now, back into town and the shop. But today I head further along the promenade and up towards Beachy Head. The Beachy Head lady was about my height (OK, a bit taller at five-foot-oneish) and her skeleton was found by archaeologists sorting out archives in a box labelled 'Beachy Head, something to do with 1956 or 1959'. She lived around 250 AD, Roman times.[15] Initial investigations suggested she was from Sub-Saharan Africa, but more recent studies think she was probably Southern European, likely from Cyprus. The facial reconstruction of her is eerie, uncanny. She was here, we walked the same ground, and more evidence that we have always been multicultural, migratory, connected. The strata in the chalk cliffs formed in the Late Cretaceous between 66 and 100 million years ago when sea levels were much higher than they are now. The cliffs themselves formed over the last few hundred thousand years by a mixture of river erosion when sea level was low, and wave erosion when sea level was high. Today, the climate crisis is accelerating erosion of those cliffs; pavement graffiti, information signs and stickers on lampposts urge us to take action now.

I think about my days in The Body Shop and what I learnt about make-up, performance and femininity. I was naive about corporations and proud of the political campaigns we were involved in. More than once a man came in shouting and threatening us because of a window display about domestic violence, and we looked out for each other when certain customers came in. Whisper networks, warning about dodgy men, diving into the office while a friend covers for you if the guy who's been pestering you 'just popped in' again. I have always used make-up and dressing up as an escape and protection. I love bright colours and patterns, and have a vivid memory of realising if folk were admiring my dress or laughing at my coat it was something I had chosen. They weren't hurting me. I also saw how important make-up and costume was to others who were discovering themselves, and that gendered violence hurts everyone. There was only one openly gay pub, and a school friend's parent worked there, so it wasn't somewhere you could really go. Clause 28 was in effect, and it is only in retrospect I admire the teachers who fought for us to read *The Color Purple*. Things were known, and not known, and queer was never used as praise. When I saw footage of the first Eastbourne Pride, and photos of progress flags outside hotels there, it moved me and I cried happy tears at my desk. In Manchester I avoided Pride weekend for many years because I found the commercialisation alienating and I'd experienced biphobia too many times to feel welcome. I gravitated to alternative celebrations instead, but Eastbourne reminded me why Pride matters.

Nando Messias' *The Sissy's Progress* (2015)[16] was the artist's response to homophobic violence they experienced in London. A dance, a mini parade, a spectacle, they walk the same streets in a fabulous dress with balloons and a marching band – a hyper-visible retort, years after and the start of their trilogy exploring what it means to be effeminate or gender non-conforming. Paul Harfleet responds to and reclaims homophobic and transphobic insults by planting pansies at the site of abuse. *The Pansy Project*[17] has planted over 300 pansies across the world since it began in 2005 and Harfleet has written a children's book about it.[18] Both Paul and Nando make hatred visible, then neutralise its damage, transforming abuse into something beautiful and infinitely more powerful.

Later, I return to the promenade because I simply cannot resist the allure of the pier. Most of it is closed, but the glass blower endures.[19] I buy a tiny fish, tiger stripes and huge googly eyes, perched on a small glass rod. I pretend to myself it has a use beyond saying, 'I was here and I am glad something familiar remains.' Inevitably, I go to the pub; the cafes are shut and the cold is sinking into my bones. A pint of Guinness and a window seat so I can vicariously luxuriate in the freezing water. I scroll through my phone and learn an old school friend has died. Years since we giggled and danced and squabbled together but still it stings. We had promised ourselves a catch-up in person but never quite did. I can see the flat where my first lover lived and where friends squatted. It's holiday apartments now, and no longer takes residents on housing benefit. I remember it as the place I was introduced to snakebite and black and the myth of romantic destruction. I'm sad and furious about the booze, and all those who we have lost to it. Sitting alone reminiscing, is this the coolest thing I have ever done, or the saddest, or just another rainy night in Eastbourne?

The hypocrisy is not lost on me. I very seldom drink on my own, I love pubs for the same reason I love streets: blethering away out of the house. The serendipity of shared space and the opportunity to chat. My preference is for a traditional boozer with a snug: comfy seats, decent chat, a good craic. Although, of course, traditionally I may not have been welcome everywhere I fancied a pint. There's a priceless roll call of cheery memories, meetings, gigs, histories, plans hatched, friendships made and treats given when I think about 'the pub'. A found poem, overflowing with tall tales: The Lamb, The Ship, The Hare and Hounds, The Eagle, The Scholar, Gullivers, The Terminus, Fringe, Sand, The Duchess, The Coach, The Grocers, The Stanley Arms, The Hideout. It's a complicated love though and I know how easily things can turn sour. I spent many wasted hours literally getting wasted when I was younger, the shame and the regret of a hangover, unwise actions, lost inhibitions, foolishness, mystery bruises and a blank memory. I'm sorry about many things I did while drunk, and sorrier still that I needed the escape of oblivion and the mask of inebriation.

This pub here on the pier is not one I visited as a teenager, couldn't get served and it was too expensive, too touristy anyway. Today, I am not dwelling on the harms booze may cause but the joy and comradeship. I get to pondering some of my many lost and abandoned projects, like *The Community Drone, Geographies of Dr Who* and *The Windowsill Project*. One of my most lamented is *Three Beers and The Truth*. It was conceived to celebrate occasional loiterer John Hawes' birthday, a chance for us to work together and enjoy the research process. It was also going to be a way of countering the flawed but frequently heard grumble that there are no 'proper' pubs left. The plan was simple: 50 pubs across Manchester, we would visit one a week or so, and share not just what we found but something of the character and social histories of the pub in question. We started early March 2020 in The Three Legs of Man in Hulme, our local at the time. Not long afterwards the pandemic closed down hospitality and as time passed *Three Beers and The Truth* faded into a dream.

When I walk these Sussex streets, there's an uncanny feeling I may bump into my own ghost. I catch glimpses of a girl, tear-stained and fuzzy, stumbling through the streets or trudging to the bus stop. An involuntary flinch at the site of an assault, a deep breath and shudder where there was an attack. There are windows I can't help but glance up at, though I know my lovers have long gone. I don't really want to see them, but I'm compelled to check. I remember walking with my nan to the shops, suggesting a detour that was less than picturesque. Eventually, she asked who lived there and I realised my secret crush was not so well hidden. Years later, I'm on the cusp of moving to Leeds. Another door, another crush, but this time there's an intercom. I'm sat on the doorstep because there was no answer and I am unsure whether they are hiding from me or if I should wait for their return. This relationship was full of the casual cruelty of two broken people clinging to wreckage from the wrong exploded star. If I could send a message to that girl with the baggy t-shirt, ripped lace skirt and scuffed DMs I would tell her not to waste time dreaming of romance. There are so many more kinds of love that matter more. I'd give her a hug and tell her not to fetishise pain, it does not make you special.

* * *

I first met Saffron Defiance Swansborough when we were in the same classes at sixth-form college. I wanted to be her friend because she looked so cool: a Goth with an excellent record collection. Of course, she was much more than that, and we became part of a friendship group I remember fondly.[20] I had some lost years in my twenties when were weren't in touch, but a chance encounter in a car park with her mum rekindled our connection which I cherish. Today, she is an educator, artist and street photographer who, after extensive travel and a career in broadcasting, has returned to live in East Sussex. Saffron documents her surroundings from oblique angles and her work frequently includes her daughters and reflections on time, memory and motherhood alongside the landscapes she loves.[21]

The last time we met the pier was closed at night so we wandered the prom together. I didn't want to spoil the evening with a tape recorder, so instead we had a Zoom chat when I was back north. I asked Saffron what she thought of Eastbourne and she said something that resonates across the UK:

> I think it is a town that is always trying to reinvent itself. And it tries quite hard to do that. And actually, a lot of the charm of the place is the structure, the heritage ... When I think about the landmarks, I think about the South Downs and that lovely gateway that we have from Holywell where you can walk up ... the sea front. The pier, and the bandstand and then down to the redoubt, I think of those places as really important parts of Eastbourne.

We talked about how beautiful the seafront is, and how the landscape we grew up in is, to quote Saffron, 'really imprinted on to my brain so ... I sort of retraced my steps consciously now when I revisit those places, I feel drawn to those places. They feel like they're part of me.' I tell Saffron I can relate strongly to this, and she goes on to describe the symbiosis when

> the person goes into the landscape, but actually the landscape is in the person as well ... (the landscape is) part of my subconscious because I've got those memories, or maybe old photographs, just family snaps of those areas, which give me another picture, like another layer or a slice of memory.

Those layers of memory that shift and unravel are something we both feel. I think of the town as a palimpsest, a series of layers, built on top of each other but the layers aren't even. Paint peels and previous coats burst through, the colour is so bright. We discuss how anachronous some of the changes feel to us, homes built on flood plains and beaches under constant threat of erosion. As residents, we saw the alleyways behind hotels and a mutual friend told us about the cockroaches in the ice cream kiosks. The attractions aren't necessarily as important to residents as the more mundane parts of town. We grew up on different estates – I was in Willingdon Trees, Saffron was in Langley. Her home was near a shopping centre I remember going to some weekends. The cheap record bin, a toy store where discombobulatingly Darth Vader signed autographs, and the electronics shop full of intriguing keyboards and radio kit.

Saffron visits the shopping centre often with her camera to document what is happening now, but also sees what happened in the past, and explains how time merges, distorts, when she looks down the lens.

I can see the current incarnation, which is the modern Costa glass. But I was there when they were ripping down the front, and I can still see it, I've got photographs of that previous part of the landscape. It was the entrance that I remember from growing up in the 70s and 80s. So I've got the two narratives going on simultaneously. In my mind, but also in my photography ... there was a Wimpy, I've got so many memories of going there. It had these tiles on [the walls] from the late 1960s or early 70s. And they lasted all the way until this new glass front was put on ... Now those tiles have been painted over with white or they've been pulled off and they're not there anymore. But they are in my head and they are in my photos ... they still exist,

I've preserved them ... it's biography, interacting with my surroundings and then that goes into the photos.

Looking at Saffron's pictures can be uncanny for me as I recognise familiar places, changed through time and viewed through her lens. Saffron only started taking pictures when she moved back to Sussex. A key catalyst was advances in technology, the camera on her mobile phone being readily accessible. We both share ethical concerns about street photography. Saffron takes care not to exploit or photograph people in moments of distress or vulnerability but has never felt it unsafe to take landscape shots. Crucially she says, 'I feel safe in the places that I visit and take pictures ..., I do feel entitled to take my space on the streets and take photographs in those places that I choose to visit.'

Saffron and I both moved to Yorkshire to go to university where we learnt feminist theory, amongst other things. I arrived in Leeds where Riot Grrl was still reverberating and I joined demos, wrote zines, danced hard and blissfully with friends. However, it was the 1990s and lad culture proliferated. Casual sexism was more than tolerated, it was enjoyed, celebrated even. It was just a laugh, everyone loves a bit of banter, don't they? Sexual harassment, violence and its threats were still there, transmogrified and amplified compared to my early experiences. To be a woman, especially a young woman, was to be a target.

Years later, in Manchester, I naively thought for a brief time that street harassment had diminished and things had changed positively. Untrue. I had merely grown older, rendering myself less attractive, or less vulnerable or most likely simply invisible in certain quarters. This is absolutely not a victory. During my thesis research our walking interviews were often disrupted by men who wanted to know what we were doing. Fiona Vera-Gray talks about harassment as an 'intrusion' or 'interruption' and I think this reflects the impact these encounters have.[22] Whether being told to 'smile luv' or 'show us yer tits', the result is a break and diversion from whatever the woman involved was doing. It's a jolt and a tear and it's impossible to know quite what is lost. We are robbed of knowing what would happen without the interjection.

The profound impact this has was articulated to by 'Sophia', a student and as I quote her here I still feel the fury and sadness I experienced the first time she told me this, shortly after she has disclosed an assault. She told me she doesn't like to go out alone anymore, and pays for a taxi home from the bar she works in if it's late, because of

> the very fragile male violence sort of thing that definitely can erupt on a night out for no good reason ... [I] get a bit annoyed at my boyfriend when he's being very not careful out because I have to live in this world where I always have to be vigilant, I always have to be careful ... because the world is so much for them [men] isn't it and they're not aware the danger that women have to negotiate ... [guys] don't realise what it's like, not feeling like public space is for you ... there are places I wouldn't go but that my male friends wouldn't even think twice about going to ... these public places half of the population is losing out ... When you plan a city you need to think are women going to feel safe here or something? It's a question that needs to be asked to bit more.

In Chapter 6 'Liverpool', I will return to the design of cities, but as 'Sophia' suggests, that alone won't change culture. The rage I feel about the misogyny we face every day bubbles up and is exhausting. I need some respite. The walk across The Downs at Beachy Head is one of my most precious in the world, although to be truthful the bus ride is perhaps even better. Catch the 702 in Old Town, opposite The Lamb and head up the hill through Meads. The first glimpse of the cliff is thrilling lush downland, bright blue sea, and the town below perfect in miniature. If you stay on, there's the delights of Cuckmere Haven and the meandering river. Today, I'll get off near the visitors' centre and find my nan's bench. It's not her name on the plaque but it's where I sit and feel closest to her. She loved this place so much we scattered her ashes here.

There's a pub over the road. I recall sitting there waiting for a friend and thinking the bar staff were the friendliest in the world. The woman who served my drinks was deeply interested in the book I was reading and complimented my clothing. I didn't realise she was trained to spot potential suicides and my solitary lingering over a pint, probably accompanied by agitation and staring out the window, marked me down as a risk. At the time, I was amused she mistook me being uncharacteristically early in this way, but on reflection I am grateful for the care. These cliffs are beautiful but treacherous and growing up here entails an understanding of the entwinement of these elements. We were taught that more people jump here than anywhere in Europe and tragedy is etched on the landscape. I used to wonder what it felt like to fly for a moment, but now I shudder about how close to the edge we got. The height is mesmerising, hypnotic even, and we would dare each other onwards in increments. We didn't understand how suddenly chalk crumbles or feet slip, but we knew instinctively this was the edge of the world and maybe sometimes its end. It's been eroding for over ten thousand years, loosing around a metre a year. Global warming and rising sea levels are expected to intensify the process, sand and shingle are moved to try and mitigate its impact. Geology again, deep time and climate change. We used to hunt for fossils on the beach below and sometimes strike even luckier with a calcite geode. I always loved sparkly things and wished I could see through every rock to find out whether there was magic at its heart.

There are legends up there too of course. The monk who appears through the fog to help you but instead lures you over. The lost souls, the vanished, the forgotten. I was told once that in Medieval times unwanted babies, the sick and elderly were tossed over to save the burden of feeding them. That stayed with me as a warning; my broken body was only tolerated sometimes and I should be glad to be alive when I was. My diagnosis has changed over the years, but many medical texts have told me my faulty genes will be eradicated in the future. The pseudo-science of eugenics seeks 'purity' and disabled people are deemed broken, worthless. It's profoundly racist and antisemitic too, and hasn't been consigned to history. Uncomfortable truths we need to know. During performance tours for the *Manchester Modernist Heroines*[23] project alongside women I admired, I raged about Marie Stopes who was the first woman to join the Science staff at Manchester University. She was also blatantly racist, wrote love letters to Hitler and did not see the working class as fully human. Women's reproductive health matters, and we all have the right to choose, but we have better champions than Stopes.

Walking the promenade in 2024 I can understand both my urge to escape and the pull to come back here. I no longer believe what a classmate told me: 'no matter what you do, here you are always the girl with callipers and a dead mum'. I've made my own way and found my own identity. I also understand now how hard walking can be, and why we need communities of care. For me, mutual support comes from friends and family (both blood related and chosen), and from my fellow loiterers. I also take nourishment and strength from the work of artists who inspire me and the women-led campaigns I have shared in this chapter. I hope The LRM can feed others in this way too. There is strength in solidarity and knowing others are facing challenges with creativity and courage, finding moments of joy in the struggle.

The sea and the cliffs and the pier are reassuringly where I left them, and I am glad to see a 'Changing Place' the gold standard for accessible toilets. However, campaign posters tell me the other conveniences are under risk and that alarms me: public toilets are an essential resource (see Chapter 6 'Liverpool'). There's traces of the 2023 Turner Prize exhibition: labels for public artworks now gone and people wondering if the space is the art. No one I spoke to felt the impact of the art world descending yet but I noticed a lot more galleries and I found one unrelated installation that filled me with absolute glee. It helped me with the treasure hunt I shared as a prompt at the beginning of this chapter. The quest for *better times* is ongoing, and the *poetry of the street* is never-ending, but just off the promenade I danced to *wild music* with *star children*.

I took my nieces to Princes Park to feed the ducks on the same lake I rowed in swan-shaped pedaloes with my grandparents. The girls' playfulness and joy was contagious. Afterwards, there was cake in the cafe and a reminder there is always something new to discover. The toilets have a big red button in them which says 'Push me. What's the Worst That Can Happen?' Doing so results in a kaleidoscope of disco lights and a blast of the Bee Gees *Stayin' Alive*. Ridiculous, exhilarating, and leaving us all boogieing our way home.

4

Stockport, Ashton-under-Lyne and Glossop

Shhhhh what can you hear? Walk in silence and listen carefully.

Stockport has always felt to me like a seaside town with no sea. There always seems to be bunting, balloons and a whiff of holidays. Maybe it's because of the sandstone that the town centre is built on. It's a sedimentary rock, laid down around 240 million years ago during the Triassic period when the area was an arid desert. Outcrops extend to Merseyside and the Wirral. The red colour comes from iron and other minerals compressed over millennia. Today, caves still lurk at the back of shops, and some were used as air raid shelters. On the outskirts are more caves, probably excavated by workers constructing the famous brick viaduct. Now they are used by people to sleep in, or they were until the local newspaper exposed them. They didn't solve homelessness with the story.

Wellington Road is the fast track through Stockport but I will always choose the scenic route. Hillgate is one of those marvellous streets where all the world seems to flit by. I first walked down it in 1998 when I got the train over from Leeds, nervous and teetering slightly in uncomfortably smart shoes. I had an interview for a job I really needed and I got lucky. I became an Information Officer for a local charity, although I had no real inkling what that might mean. It sounded interesting and useful and was enough for me to cross the Pennines.

I was told later I was the only person who applied that could (just about) understand desk top publishing.

I worked in Russell Morley House, named after a man who in local myth was a James Bond type and the first male to join the Women's Institute. We were opposite Robinsons Brewery which still chose to employ Shire horses to deliver the beer we could often smell brewing. Stockport was the scent of malt and wet stone, with its southern boundary marked by wafts of chocolate from the McVitie's biscuit factory. In the immediate vicinity to my office were second-hand record shops – one with gramophones in the window, the other cheap CDs. There was a fantastic Gothic Emporium, I was sorry when it closed, but also glad to get the black PVC trousers with holographic stripes which I had long coveted cheaply. They squeaked every time I walked in them. There were also several 'private' shops selling pornography and sex toys. One day, a local dignitary came into my office to complain that the British Red Cross logo was being misused on a fetish costume. Pound shops spilled out onto the street, and further down The Underbank were boozers, chippies and buttie bars leading to the vision of a future that never quite was, the Merseyway Shopping Centre. The Woolworths was handy, and on the side of British Home Stores stands a stunning concrete mural depicting key figures in the town's history. The mural is still there for now, but Woolworths closed in 2009 and BHS went into administration in 2016.

Arriving by train one sunny day in 2024, the town's transformation is dazzling: a glamour in all meanings. There's a plaza which fits the contemporary pattern book and feels a bit hollow, a bit soulless, and for a moment I am confused about where to go. I can't retrace a journey when the streets are gone. However, the new bus station nearby makes it hard to be churlish. I have no qualms about the better lighting, seating and roof garden. When I stopped commuting by train from

Leeds the 192 bus was much cheaper, but the bus station was never alluring. Sometimes I'd walk down the road to a less convenient location with a less intimidating atmosphere. Kids would often throw stones and bottles at the bus window, and it was a very long journey to be sat next to someone who you really didn't have the energy to deal with. I think I would feel more comfortable waiting here now, but I'm less convinced the journey itself would be any more pleasant.

Back down to Hillgate in 2024 and the street is mid metamorphosis. You can almost smell the entrepreneurial frontiers of a town dubbed 'The New Berlin'. It was a hyperbolic quip from a Manchester DJ, but it's been seized on by many. It's a reminder although the evolution of creative places often has organic elements, the development of a district is rarely an accident and economic systems are never natural. There's a parallel to Manchester's Northern Quarter discussed in Chapter 1. It is true creative industries, artists and makers need cheap places to rehearse and work. They will congregate where those kinds of places are, and in doing so, if they are lucky, create a community, and a scene. These are Good Things. There will also be investment opportunities, place-making strategies, incentives to relocate and regenerate. It was a 1993 Manchester City Council regeneration study that galvanised the process in the Northern Quarter. Strategies included commissioning public art and self-consciously quirky features like bespoke street signs and ceramic bird sculptures perched on windowsills. I'm not snarking or being cynical here but I am calling time on the idea that this all just kind of happens. The success of the Northern Quarter now prices out new entrepreneurs or artists without significant capital – but gentrification is not the fault of artists. Affordable housing and studios would boost everyone, giving working-class creatives time and space to experiment, dream and make what they want. In the meantime, there are places like Stockport which have not been saturated yet and where newcomers can join what was an already vibrant place.

I think about the power of policy and money as I sit in the window of a vegan bakery a few doors down from Russell Morley. The coffee is excellent and so is the cake. I chat to the owner who is on first name terms with most of his clientele. We talk about the changes. One of the few constants is opposite. The tattoo parlour where I got my nose pierced one lunchtime. I think the old gramophone store is another coffee shop and bar now, but my memory is a bit jumbled. There's an enticing looking bookshop where I bump into a former colleague. He is doing some research on Edgely Park, and I tell him about my friend Rose Walker. She was a trustee at the charity I worked for and an absolute powerhouse of kindness, compassion and getting things done. The community centre she used to manage now bears her name and I'm delighted she has a living memorial. So much more fitting than a plaque or statue. Russell Morley House is conspicuous as the largest boarded-up property. I've been told there were structural issues that mean it's not a great investment, as redevelopment advances, I suspect it's only a matter of time before it's demolished. My office window is cracked, too small, high and insignificant for the boards. The ground-level ones are painted with fruit and vegetables. The adjoining space was once the Pop-In Cafe run by Age Concern. Cheap meals, hearty portions: pea and ham soup, roast chicken dinners, steak and kidney puddings, winberry pies. Mostly it was a gathering place, the social function as nutritious as the food. I would sometimes venture down the rickety back corridor to talk to the cook about music and scone making.

Food, cooking, sharing and eating it have always been central to community building and it's perhaps no surprise that walking artists have drawn upon this. Clare Qualmann's *East End Jam*[1] is a great example of how walking artists build food into their work. It's a long durational project based in Stratford, her London community. It begins with a walk, or several walks, to forage for wild food. Clare started the project after some unexpectedly fruitful walks in her neighbourhood. The website says,

> On the ordinary light industrial and residential streets of the city I found plums, elderberries, sloes, apples and blackberries in abundance. I picked these and made a range of jams, then led a walk entitled 'Stratford Jam' showing people where the fruit trees and bushes could be found, and gave them the jams to eat.

East End Jam has evolved since then, and there have been various iterations and community jam cooking sessions. The recipes are shared on the website.

East End Jam is typical of Clare's work, blending art with the everyday and rooted firmly in her lived experience. She said to me she had thought a lot about whether this is walking art or not, and concluded that it is:

> I did ask myself is it Walking Art? Yes, because people come on a walk which is at its core. For me the walking [rather than any documentation] has to be the art, and sometimes walking is a method to create, sometimes it is exploring a theme or following a trajectory which has its own lurch … East End Jam is about the environment and food as well, not just walking, they all align in East End Jam. The work is the method and theme and subject, sometimes walking is a vehicle for something else.

As Clare said, as her practice developed she started to think of the walk itself as a creative space, exploring the shifting space between home and work and the environmental crisis. I tried to think about how The LRM could encourage people to taste the city safely and for free on our walks, but our foraging has always been a bit ad hoc. We have thought about the *Lickable Cities*[2] team, some of their members have walked with us, and their absurdist research makes me smile. However, their lick everything approach felt too risky for a First Sunday.

That very question was inconceivable to me when I was working in Stockport, a time before I ever dreamt about walking as art or that I could be an artist. I was too busy writing newsletters and guides, delivering training, sharing information. My role was supporting community groups, voluntary organisations and charities across the town. I discovered I enjoyed being part of the supporting infrastructure, it gave me an aerial view of the action, and it meant I could feel like I was contributing support to a range of causes without having to choose. Good community development work should be invisible and not remembered, dissolved into what an organisation wants to do and the skills it has gained to do it. I became aware of the many practical ways people work together to make change happen and care for each other. I also became critical of charity and patronage: mutual aid can be so much more effective and sustainable.

It was delightful, but actually I often wasn't there in Russell Morley. I got to go and visit groups, exploring across the town. Stockport is a patchwork of very different communities. That wasn't my only absence though. For the first time, I was fully connected to the internet. Lonely, trying to find my way in a new city, trying to deal with bullies and coercive control (actually that's not quite true; I had not really recognised them as such yet). I discovered another world I could escape to.

I roamed across cyberspace making connections with people who shared my interests (mostly musical). Many of the friendships made then still endure, although the platforms we met on are long gone. That makes me feel ancient, as does seeing clothes I wore back then in the new vintage stores in Stockport. I'm of the last generation not born into a fully digital world. I can remember the thrill of discovering the internet and all the amazing possibilities it held. Cyberspace is another commons where we need to be both wary but open to potential. I don't want to gloss over the harms of online bullying or social media abuse, especially the toxic masculinity and hate that festers so awfully. Neither do I want to ignore harassment by direct messaging or the increasingly alarming political influence of tech bros. These are serious harms. But here and now I want to focus on the positives of online connections.

I think about this technological capacity for bringing pleasure when I walk with Blake Morris[3] during his *British Summer Time* walks. Blake has been in Northampton, New York and Mexico City when I have participated. The idea is to walk for 15 minutes prior to sunrise every morning for the week before clocks go forward or back for British Summer Time. He is joined by folk all over the world, and when I have been able to participate it's been fantastic to get out of bed and watch the light change. It's a part of Blake's wider creative practice which explores walking and connection. On a First Sunday last year, we explored Salford Quays, an ostensibly spectacular, sanitised area of art centres and media production studios. On the edges are immigration detention centres and stalled developments. We met on Broadway Metrolink platform, while Blake was in Brooklyn Heights. We shared pictures, videos and audio recordings and a video call at the end. Blake and I have walked together in person too, meeting in London and finishing at a bookstore selling remnants of Doreen Massey's personal library. This prompted an exchange which we published where we talk about what might make walking better and more accessible in the future.[4]

Distant walking, or walking together online, became a valuable source of connection and comfort during the Covid-19 pandemic when there were legal restrictions on where, how often, and who with, one could walk. I'm deeply ambivalent about First Sundays' transition to WhatsApp.[5] I was too stubborn to let anything dissipate the pattern, but often I found the experience distressingly hollow. Sonia Overall's[6] *Distance Drifts* were an exemplary case of how the restrictions inspired new work from walking artists. Sonia's *'Drift Deck'* was created pre-pandemic, a deck of cards each of which contained a creative walking prompt. She told me this was 'things I was doing in my teaching quite a lot, I found myself constantly generating prompts on cards and slips of paper … thought I should refine this … for other people to use'.

Sonia was contacted by people who asked if they could use the *Drift Deck* on Twitter during the first UK lockdown; she said yes and offered to deal the cards. She remembers:

> We were all just stuck in back gardens and local parks and some people were locked down indoors so that's what we did, we agreed on a time on a Sunday morning and I just dealt some cards, took some photos and everyone said oh that was fun shall we do it again next week? Then I thought actually shall we just keep doing this?

Distance Drift took place weekly for four years, and then became monthly. It's beloved by many. At the time of writing, it is still active, with Sonia facilitating at 10 a.m. on the first Sunday of every month now on Blue Sky.[7] Reflecting on the project she says,

> it was quite a task I set myself, something new every week, so I'm slightly relieved … but there's a lot (of prompts) in the bank now people can still walk… and it's been such a lovely thing, I've met, networked virtually with, so many

people because of that project which has been lovely and found so many like-minded souls.

Incidentally, Sonia Overall proudly claims the title of 'psychogeographer' despite facing criticism because 'I do feel strongly that's wat I want to say because it does exactly what it says on the tin and I don't see why I shouldn't be able to access that title.' Sonia is one of the current organisers of the Fourth World Congress of Psychogeography and goes on to say 'I think we are all natural psychogeographers, my experience bringing up my son was he just intrinsically ... didn't need prompting, just needed permission and he was acting as a psychogeographer.'

* * *

I left my job in Stockport in 2003. I was revived and reinvented, changing my name and having a celebratory naming ceremony with all my loved ones. Knowing you can choose your own name, and exercising that right, is a splendid feeling. I travelled across the USA, often staying with people I had never met in person but were part of my online communities. When I returned, I got a similar job in Ashton-under-Lyne, another Greater Manchester satellite town with a proud identity and a market at its heart. I could buy a carrot or apple or a handful of cherries in a paper bag and didn't have to plan meals for a week or organise a visit to a big supermarket, which I found very pleasing. It was while working there I became involved with The Basement, learnt about anarchism and the SI, and The LRM was formed.

I was in the Ashton office when news came through of the fire in the Northern Quarter that ended our Basement dreams. Devastating text messages: 'turn on the TV! I think the gaff is on fire.' My manager told me to go, called it compassionate leave, they knew I could not concentrate. I found myself standing in the middle of the road at Stevenson Square. This had been a rallying point for the Suffragettes and later a traffic snarl-up; the long mooted pedestrianisation finally began during the pandemic although it revolved around beer gardens and cafe tables. There, in 2007, I stood staring, helpless, so insubstantial the wind blew through me, my desolation punctuated with hugs from friends and pestering from journalists. One of them lied to convince me to talk to them; when I returned to work, women in the sandwich shop were excited to have seen someone they knew on TV, they didn't register I was tear stained and shell shocked. Ironically, most of the damage to The Basement was caused by sprinklers rather than fire – but the building as a whole was unsafe and we were locked out bar a few salvage trips. I'm not going to pretend our anarchist utopia was perfect: maintaining the centre was a lot of hard work, and many of my memories are of washing up, cleaning toilets and endless consensus-seeking meetings. Access was also a big problem; we were downstairs and naively assumed installing a lift would be easy if we held a few fundraisers. Planning regulations and licensing conditions were unexpected challenges. But I will always cherish the camaraderie and sense that we could change the world. We were confident that somehow, we were making a positive difference in the world and I think we did.

I left my job in Ashton not long after the fire, I needed a role with more hours and less commuting. When I finished my stint there the new Ikea was open, and we would get dosas from Lilly's before taking the bus home. The tram would come later, the bus journey took me through the changing landscape of East Manchester. The stadium was built for the Commonwealth Games, now it's home to Manchester City football club. Local campaigners continue to raise questions about finance from the United Arab Emirates. The B of the

Bang statue has long gone, its spikes too dangerous. It feels a long way from Ashton. Ashton still remains a site of resistance, rebellion and care as exemplified by Pauline Town at the Station Hotel. Her pub is a sanctuary for her town's homeless people, offering food, friendship and a pointer to support services. It's also the epicentre of *We Shall Overcome*, a coalition of musicians and artists providing 'a raised fist and a helping hand'.[8]

In Manchester, I worked in a converted church overlooking Ardwick Green. The eponymous park was one of the first to be owned by the local authority. It had been managed by, and for, the wealthy suburbanites but they didn't want the responsibility. There's a large rock in the park with space for a plaque that has long since fallen off or, most likely, been taken as a souvenir. I worked on a social history of the area, partly inspired by reading about a tour guide who refused to step foot there, dismissed the local community. Snobbery. And ignoring the fascinating stories of the neighbourhood, including the formation of 'Voluntary Sector Row', home to many charitable organisations. They included George House Trust, which evolved from the volunteer-run Manchester AIDSLine. There are dolphin motifs on nearby fences, allegedly a protest about taxes the wealthy once had to pay. The rock in the park is a glacial erratic, green slate from the Lake District, probably deposited in the area around 25,000 years ago by a kilometre-thick ice sheet. Local school children have been told it's a meteorite, an alien being who made an amazing journey across space. Two incredible journeys – does it matter if they learn the wrong story? If a meteor had landed there the crater would resemble Arizona more than Ardwick and we would be staring into a big hole.

So many of the organisations I have worked with have closed due to cuts. Infrastructure isn't valued, but I can understand why in times of austerity. Frontline work is more urgent, but thriving is more than survival. We collectively need to dream, and plan, for better times when charity is not needed. The arts are a vital part of that. When I was talking to Dee Heddon[9] about her work, she told me 'theatre is a social practice, my love of theatre is it's a way to have dialogue, discussion about things and also a way to show different worlds, like what if the world was different? How can we make it so?' Dee became interested in the field of walking art when she was writing about performance and autobiography and realised most of the art she found was by men. Dee hadn't intended to make walking such a focus of her career, but says things evolved from her *40 Walks* project. This had been a gift to herself for her 40th birthday, 40 walks with 40 people because she thought it would be 'lovely just to slow down [spend] meaningful times with my friends'. The resulting blog inspired many.

As we've seen, walking is also social, and a way to imagine better worlds. Dee asks, 'What can we learn from those artists, women artists that change narratives and knowledges, really changes knowledge about what walking art is and dominant scripts of the heroic male solo walker?' I think of the times I have been lost in the ecstasy of art, and how that energy can sustain us and bring people together. It may not be forever, but the ideas can linger. We need to be able to dream new futures into being possible. Environmental collapse and the climate crisis are key areas where we need to think about the future and embrace collective responsibility.

Some artists have done this through epic walks. Monique Besten walks across Europe, usually alone, camping wild, often collecting rubbish and engaging with environmental themes. As she travels, she stitches stories on her clothes to create a lasting legacy. She says of her *Soft Amor* walks:

It didn't start as an activist act, a political statement or social or cultural research. To some extent it has turned into that.

Not because it is what I wanted or aspired but because that is what the world wanted. I just walked. I walked for weeks, months, and people on the road asked me questions, told me their stories, gave me their opinions. They saw things in my walking I hadn't seen myself. They hardly ever asked me if it was art. It didn't matter. I loved that. It was so different from interacting with an audience in a traditional art space.[10]

Jess Allen[11] has made walking artworks across different scales and durations. *Trans-Missions* followed rights of way that were as close to the line of electricity pylons as possible. She said,

The rules of the performance score dictated that I must speak to or otherwise acknowledge every person I encountered, forming an unbroken line of transmission. Carrying a rucksack of 50 low-energy lightbulbs, I gave one to each stranger who invited me into conversation and who offered me a message to pass on – to transmit – to the next person I met. This message was requested in exchange for, or response to, the message I passed on from the last person I had similarly happened upon.

Each lightbulb Jess dispensed carried an invitation to a Midsummer gathering. Answering Dee's call, both Monique and Jess have story sharing at the heart of the works discussed above. They interact with people they encounter on their walks, imparting knowledge and modelling new ways to move through the environment. They don't leave a physical trace but instead plant seeds with their stories. In New York, Eve Mosher's walk left a tangible mark that signalled to a possible future unless we act now. *High Waterline*[12] saw her walk through the city drawing a blue chalk line to indicate areas that will be lost to climate-related flooding. This very simple but effective tactic inspired similar walks in many other places.

Impactful walking does not have to be grand in scale. Shonagh Short[13] curated a WalkCreate project 'To *The Moon and Back*'. She wanted to call attention to the walk to school, a profound act of care and love. She worked with parents and carers from the school her own children attend, learning the landmarks each come to know. These could be trees, or places with puddles to jump in, or cars that seemed to be abandoned. Local features, but also resonant across other schools in other places. Adding up the distance travelled would make an epic journey in conventional terms, but could not begin to account for the relationship building, love and care poured into the trip. She asks:

> does it have no status because of who we are, because we give it no status? Does it have no status because it's not exploration, it's a repetition? It's the same route over and over again, an unimportant, unremarkable thing that is repeated ten times a week, 42 times a year, over seven years or ten years, however many years, depending on how many kids you have got.

Across the country campaigners for school streets, safe routes for walking and cycling, and play streets call for a similar reorientation. They ask why must cars dominate, what if we could transform streets into something, somewhere else for playing?

Roads, and traffic, are by the far the biggest complaint in many places I have walked and worked in. In 2022, I was one of the artists in residence at George Street Bookshop in Glossop, Derbyshire.[14] It is community owned and co-operatively run after a buyout when the previous owners retired. I was there to create a *People's Tour* and spent time collecting stories. I had planned to create a walk visiting favourite places and locations connected to those stories, but like the best participatory projects it did not end up like that. The things people loved about Glossop were intangible: the horizon, the feeling of community, the light in the trees. The specific places that were mentioned were generally very personal, the homes of loved ones. However, when asked for least favourite aspects, the answer was unanimous – always the traffic.

There have been ongoing arguments about a by-pass for many years. The local heritage trust document 2000 years of traffic in the area, since the hills above were the site of Melandra, a Roman fort.[15] Centuries later, there were salt paths as the precious commodity was transported north from the mines. Then the pure brooks and climate made it a centre for cotton spinning. It enjoyed the patronage and support of the Howard family, Duke of Norfolk, who built many of the civil amenities. Historically, it was much less radical than its neighbours. It was Luddites from Ashton who closed factories in Glossop, and the Chartists were not so active as in other areas. Today, the town has a reputation for being socially and politically conscious, there's an active Labour club and many local environmental groups concerned with issues, including the Mottram by-pass. It has been discussed for 50 years, and campaigners are not happy with the chosen route, but work is due to start in 2025.

There's another conversation bubbling under in Glossop too. Local business owners talked about the impact of the pandemic, and volunteers were concerned by the increased demand on local foodbanks and other support services. The market hall has been closed for renovation for several years and the reopening has been delayed again. Familiar concerns about increased rent pushing stallholders out are being voiced. Visiting a local arts group, people told me how their children were being priced out of the area. The heritage centre was long closed but there was a strong desire for it to be reinstated. Fears of creeping gentrification and a loss of richness and character were being voiced, hipcholia on a different stage.

Glossop is on the edge in so many ways, and I hope the conversations about change are public and make space for all voices. It is physically now at the end of a railway line, with the connection to Sheffield severed by Beeching. The valley terminus was often seen as a benefit, an incubator, a reason for Glossop's distinct character – but also meant some felt cut off, ignored or left behind. Administratively, Glossop is in the High Peak, but depending on the form to be filled it can be categorised as the East Midlands, Derbyshire or Greater Manchester. The local hospitals for residents are Tameside, Wythenshawe and Stepping Hill in three different boroughs of Greater Manchester. Tameside was created in 1974 under the Local Government Act. There are plenty in Greater Manchester who still feel aggrieved, I was quietly surprised how little celebration of the 50th anniversary of the ten boroughs there was. Then I remembered my first week in Stockport and how many people insisted on an address in Cheshire, and later meeting campaigners in Oldham adamant they wanted a return to historic counties. Identity can still be linked strongly to places and names do matter to people. Words, and labels, have an impact. This debate is very alive in Glossop, with some wishing to be amalgamated into Greater Manchester and others fiercely defending their local identity. The social media debate has been quite unpleasant and almost jingoistic, but in person it is much more amiable.

Even the green spaces have multiple names. At the end of George Street, just a few minutes away from the shop, is a bridge over Glossop Brook. Just before you get there, a patch of woodland has been annexed and covered in no entry signs, a private micro-enclosure which is heavily contested locally. Over the bridge it is very much open to all; there's several paths here, I prefer the one which runs alongside the brook. The area is known to some as Sandholes, others as the People's Park, Harehills, Pinch-Belly Park or George St Woods. It's a gentler debate, and the park is well used by all. Local women told me how different it felt to them than the city, it was much safer and not too wild. They could wander in this space quite safely (though most also told me they took standard precautions such as a fully charged phone). A woman who didn't give her name told me she came here to sing to the trees when she needed to clear her head. My friend Caroline Turner goes litter picking along the brook, and another, Joan Rutherford, told me about the community garden planted along one edge. It's just past one of the few pieces of graffiti I saw in the town, a line from *Desiderata* by Max Ehrmann: 'go placidly amid the noise and the haste, and remember what peace there may be in silence'. I hadn't thought of the poem since my time in Stockport where it was pinned to the wall by the photocopier.

Soundwalks are an increasingly popular mode for walking artists. *Sound Walk September*, an initiative from online network *Walk Listen Create*, uses this definition:[16] 'A sound walk, or walking piece, is any walk that focuses on listening to the environment, with, or without, use of technology, or adds to the experience through the use of sound or voice. This can include a scripted or choreographed score or work that has additional audio elements.'

A listening walk can be a profoundly moving experience. The LRM have walked in silence together and found it amazing how many textures and dimensions to sound there are, how swiftly the soundscape changes as you move through the city, and how engaging through close listening opens you up to a new appreciation of the sonic realm, its possibilities and its impact.

Audio can be used to guide, disorientate or augment your walking. Zarina Dolan's *Running Shoes* is discussed in Chapter

3 'Eastbourne'. This wasn't initially conceived of as an audio walk but she adapted to the restrictions of Covid-19. Zarina Dolan said: 'I had never considered the use of walking as a creative resource until all of the other common ways I made theatre were stripped away as a result of Covid-19 restrictions. Having to redefine what theatre meant to me and where theatre can take place has really broadened my creative process and I thoroughly enjoy using walking in my process now.'[17]

Dance artist and choreographer Laura Fisher produced two audio performances in response to Covid-19 with music by Sonia Killmann.[18] *GOING OUT/GOING IN* is designed for headphones and a walk in urban space to 'contemplate the built environments we inhabit and the ways they move us'. *GOING IN/GOING OUT* is a broadcast or download for audio speakers to be shared and 'contemplates the body as a site for memory and bodies as a shared site for choreography across time and place'. Laura explains why they were inspired to make the work as a positive action in difficult times.

> As a disabled person, I have experienced long periods of being unable to leave my home ... So the restrictions of lockdown were not particularly new to me ... From being in conversation with other people online, I recognised there seemed to be a common struggle among many people in adapting to and coping with the confines of this reduced world. Through my previous experiences of isolation, I had developed techniques for finding new ways to notice and experience the environments around me ... [I wanted to] create a participatory performance which guided partici-pants through some of these ways of thinking and noticing, as an offering and gesture of love to other people in a time of crisis.[19]

Laura implicitly understood that walking is not accessible to all and so devised 'sister walking performances, which offered similar practices and techniques for re-experiencing the familiar and connecting internal and external landscapes'.

Alisa Oleva and Debbie Kent walk and work together as the *Demolition Project*.[20] They have created a range of sound walks using lots of different tactics. Regular group walks invite participants to listen intently to 'urban uncertainty' at building sites and in the middle of regeneration projects. When I spoke to Alisa she said, 'we would walk together even if no one comes, we just do it'. *The Demolition Project* has also made a number of site-specific sound collages and audio pieces which evoke movement, change and absence. For example, *The Street: a User's Manual* is an audiowalk that turns the streets of Bow into the performance with a score that suggests different ways to move through and experience the city. Their collaborative practice took its name from their first work together. They invited people to tear something from a map – to demolish it and leave a note explaining why, so the map has holes full of many new stories. Alisa tells me they sometimes walked to where the hole was, but that this wasn't walking art per se. However, she feels it has all the elements of her later work and embodies what is important in her creative practice, 'giving agency, empowering people to make decisions, bringing personal stories, bottom-up archives'. Modifying the map is challenging the accepted narratives and sites of power. 'It's about connection to places. It's about people's stories and how we connect the city … Walking is very present for me in that work.'

Alisa[21] says she holds the title walking artist proudly and relishes 'all the confusion and questions' when she introduces herself as such. She describes her practice best by drawing parallels across art forms: 'a painter used paint and canvas, walking is my material and the city is my studio'. Alisa emphasises the freedom, mutability and power a walking artist can exercise without the need for patronage or expensive equipment: 'you don't have anything, and you can do everything by just changing how we look at the world ... very low key but with potential to create alternative stories'. Her words very much resonate with the spirit in which I walk with The LRM.

Alisa lived in Moscow as a child and moved to London as a teenager. Many of her walking art projects have been international, often focused on Eastern Europe. Themes of migration and gender reoccur across her work, although they were not always intended. As she reminded me: 'walking as an art form, it always starts from your personal experiences, how you navigate space ... [it] puts the personal is political into a wider context'. When I asked her what she felt most resonated with this book, she suggested *Walking Home*[22]. This project was conceived as a residency in Istanbul, the intention being to meet women at an arts centre, then walk one-to-one with them wherever they chose when she asked them to 'walk me home'. The in-person residency was cancelled due to the pandemic, but Alisa still wanted to complete the work.

She adapted *Walking Home* to offer a 'space to be together differently'. Alisa had already developed the vocabulary and tools for a remote residency because she was used to connecting with people she could not physically be with. In this instance, telephones were used; and Alisa and each participant would talk to each other as they walked. In Istanbul, there were multiple starting points to choose from, each linked to women's history. This element and the research needed for it were not present in the original plan, but the map remains as a legacy. Whichever starting point participants chose, they were met there by someone from Perform Istanbul who ensured they felt safe and comfortable.

Alisa told me she found the whole experience 'really beautiful'. Thematically, the meaning of 'home' became heightened or shifted for many women during pandemic restrictions. The implications of 'stay at home, stay safe' were not equal for everyone. Alisa was overwhelmed by how many women responded to her invitation, and thought this was because of an intensified desire, and limited opportunities, to experience live art and performances, to 'step out of the usual'. Significantly, the home many women walked Alisa to was not necessarily where the participant lived, but somewhere they felt welcome, safe, at home in the city. Alisa reflected, 'it's a shame I didn't go to Istanbul, but in a way I don't regret it'. In some ways, she felt the remote experience was 'much more deep' and many more women could participate as Alisa was not limited by the period of time she could stay in Istanbul.

The Demolition Project used maps as a way to collect untold stories about urban space. The LRM *Cake Map* tried to do something similar. We made several hundred cupcakes, and each was topped with an image of a building in Manchester. People were asked to choose a building to devour with love or consume with hate and tell us why. We swapped their stories for cake. It was not topographically accurate cartography, but both times we created the installation it was a lot of fun and we heard some fascinating stories about places people love and hate. Our motivation had been to challenge conventional consultation methods which we felt rarely engaged thoughtfully enough with communities. The LRM *Cake Map* was also a stealth vegan outreach project; we wanted to prove how delicious our baking could be, and maybe gently challenge some assumptions. It was truly collaborative, with loiterers baking and sharing cakes, the pleasure of swapping recipes, community building by food. It had been a daydream of mine for a while, and to see it realised felt very special to me. Initially, I planned to base a walk around the stories we collected, but realised the range, and their geographical spread, made that tricky. Instead, they filtered through and cross-pollinated many other walks. This felt a similar dilemma to the one I faced in Glossop, coincidentally home to Caroline Turner, founder of The Cake Liberation Front, who contributed a lot to the cake map's success.

My own relationship to Glossop is detached enough that the names people use for a place are charming to hear, but I don't have a preference. The tour I created explored what they might imply, and how those lines on a map, those connections across and beyond them that Doreen Massey articulates, might change how we feel. For me, Glossop has always been a place of love and friendship and I shared this as part of my performance on the tour. I'm not what you would call conventionally romantic, Glossop is the only place I've had a Valentine's meal. It was served by my then lover's mother, back when I lived in Leeds and a free meal was the most welcome kind. She gave me a teapot which I cherished long after the relationship floundered. Later, my Glossop was a place for festivals and celebration, often again connected to food. The Globe, a traditional pub that has been fully vegan since 2002, was the end point for many rambles. Friends moved here, organised gigs and parties and day trips. And, for reasons unclear after all this time, it's where a deep platonic kinship, queer and joyful and cruelly intense, ended. A summer laughing, a winter dancing, they taught me the nature of more birds, plants and feelings than I ever dreamt could exist. I sobbed in Norfolk Square because we had a chat and I knew it was the end. Insubstantial on the train home, each jolt threatened to blow me away. Disintegrating at Flowery Fields. A wraith at Piccadilly. My world reconfigured on sand and ash and glitter. Years later, the curse of the pre-booked train seat and we exchanged small talk, shared packed lunches. I kept breathing, felt nothing when they alighted, and I carried on to Glasgow, a city that always revives my spirit. We don't have enough words for love because still some folk are shocked a friend can break your heart more thoroughly than a fuck.

My residency at George Street Bookshop underlined the relationship between walking and writing. This connection is further explored in *The Walking Library* by Misha Myers and Dee Heddon. They ask, 'What Book would you take for a walk?' The first iteration of this work was for Sideways Festival where they walked with rucksacks full of 90 books on walking and invited people to join them. Their website explains what this means:

> Walking with a library of books, we wonder what these literary companions add to the journey; how collective reading and writing in situ affects the experience of the journey, the landscape and the experience of walking; how journeying and the landscape affects the experience of reading; how reading affects the experience of writing; and how a walk, as a space of knowledge production, is written and read. Through walking, reading and writing together we create an immersive and moving space, a kind of mobile laboratory.[28]

The library is more than a collection of books, when participants choose to take a book out they are invited to share why. They walk as a 'peripatetic reading group'. I have participated a couple of times, and was beguiled. It was fantastic how the group coalesced so easily, bonding over a love for books with extracts which seemed to perfectly illuminate the moment. When I spoke to Dee, she explained how the walking library was designed to be collaborative on a number of scales

1) It is a collective project between Dee and Misha
2) Participants on each walk choose a book and how actively they want to share (no one is forced to talk)

3) The collection of books in each library includes crowd sourced titles and suggestions. If a book is donated, or acquired because of a suggestion, a card with the reason why is left in the book – an 'autobibliography'
4) Different editions of the *Walking Library* have been commissioned, so the particular focus of each library is in itself collaborative.

This multifaceted collaboration is a theme throughout so much of the work discussed here. Walking Libraries have appeared at a range of international events including *Walking Libraries for Women Walking* (2016–18) in London, Edinburgh and Bristol, UK, and Geelong, Australia. The catalogue of books is available online, together with fieldnotes from all the iterations.

I very seldom document LRM walks. I never wanted to stop people exploring themselves, and I worry that being too prescriptive may hinder serendipity and stifle immersion or spontaneity. LRM prompts are not meant to be didactic and a definitive route would be anathema. Sometimes I take pictures, aid memoirs, and fellow loiterers do too. Occasionally, there is some collaborative documentation if we have had a guest facilitator. An example would be when we were joined by Isabella Pojuner. She was part of the *Narrow Margins*[27] project, investigating the criminalisation of trespass in England and Wales. This vital work interrogates changes to legal definitions of property and trespass in the UK, paying particular attention to the impact on marginalised communities. They site the acceleration of trespass laws in historical and colonial processes which continue to cause harm. This was the invitation she extended to us. It chimed with many of our recurring themes, but we recorded it in a different way thanks to her invitation.

I research trespass in my job and practice mental sketch mapping, a form of critical/creative cartography. In both I aim to unlearn property logics. This is needed in order to build property and memory cultures that are engaged with ongoing dispossession in Britain, while remaining in conversation with settler colonial and imperial projects the British state has conducted elsewhere. To this end I have been mapping trespass and access signs everywhere I walk. This Sunday I invite you to join me in mapping signs in Manchester. This will be followed by a mental sketch mapping exercise (I'll lead you through what this means when I facilitate) – please bring plain paper/plain note-books and a pencil/pen for this.

Mischievously, we began at a No Loitering sign painted on a wall off Oxford Road. When we sat down to draw our maps later they were fully of bossy directives and diversions. 'Footpath Closed'. 'Park Gates Will Be Locked at Dusk'. 'Diversion in Place'. 'Caution'. Arrows pointing pedestrians in seemingly conflicting directions. 'CCTV is in Operation for Your Safety'. We also created a communal poem after December 2023's First Sunday. It shares what we found on a drift guided by Phil Smith's book *The Silversnake Project*[26] Our meeting point had been the glacial erratic in Ardwick Green.

A December Dérive
We became a river
Whirled together
Trapped by metal
Funnelled by traffic
Moving forces beyond us
Cut off from our life spring
Crossing roads as impossible as parting the sea
And and and and yet

We drifted as one
Followed the signs
Subtle, tender, shifting
Marine life washed ashore
Meandering
Maritime beyond time
Invisible waters leaking through
Dolphins surviving despite decay
Riversides with no river
Desecrated Medlock
Tenticular forms crept apace
Ghost kitchens, dark brews
Lovecraft
Love. Craft.
Shipwrecks
Lucky beds
Sacred groves
Treasure troves
Labyrinths
Demon dogs
Glass fish for drunken sailors
You couldn't make this up
Together we knew without knowing
Tides guided us home
A toast to terroir,
To aqua,
To stranded mermaids,
Wailing foghorns,
Celestial lighthouses,
Flotsam and jetsam rising again,
Cloudwater bubbles,
Cliff edge crumbles,
Two hearts, eight legs,
Multitudes

No borders
No cares
A black star on a red button
Something to cling to as we float on by

During the Covid-19 pandemic, sociologist Emma Jackson and colleagues at the Centre for Urban and Community Research (CURC) Goldsmiths convened an online workshop exploring walking and writing. There were practical elements, and this was a record of my lunchtime walk that day. It forms part of their edited collection *Writing Walking (One day in Late Spring during a Global Pandemic)*.[25]

Pass the carwash and the church,
Zig zag across road works
Smell weed, weeds, smoke, barbeque, tarmac
Teleports me back, 1986ish
Sticky pavements, newly covered, Uncle's new polaroid camera
Playing out paused for an awkward family pose
Not a special day but it's stuck, the only photo I have of us all
Snapped back by a phone, gravel crunches, car swerves
In the park (is it a park? It's just a triangle with benches and bushes)
in the maybe park an encounter not meant for me
a secret exchange I don't want to interrupt
hoods up all round
veer right, past garden statues: Buddha, lion, windmill, gnome,
Painters scaffold, waterproof bright
A flurry of blue-tits, not the right word but the right bird and they delight me
Onto the main road and the pub.

Closed of course
I peek in the window, stools on tables, Christmas menus,
    clear bar
I miss it pangs
Lovers rock on Friday, slowly speeding up with every one
    for the road
Stolen afternoons shouting at gameshows or mending a
    world
Guinness in a glass, the meaningless, meaningful blether
    with the regulars
I wish The Three Legs back to life
And wonder
How will we mind the gap?

I know anyone else would have focused on different aspects of the landscape, and doubtless made different connections. I didn't get to walk in person with Emma until 2024, when we were also joined by Louise Rondel. They had been walking, mapping and listening to The River Quaggy in Lewisham,[24] exploring what the river meant to communities along it. The soundscape was pivotal to this work. Together we walked the River Irwell around Salford Quays, and I marvelled how varied the sonic environment was. I had assumed there wasn't much scope for different noises here but, when I listened carefully, each bridge we crossed vibrated in a different way. Circling an office block every side had a subtly different aural texture. The Metrolink announcements cut across bird song and the bustle of outdoor cafes. The snatches of conversation we caught as we drifted by were a tantalising mash-up of the passionate and prosaic, all seemingly polite and uncensured. There had been a PSPO in the neighbourhood which banned 'foul and abusive' language. Comedian Mark Thomas held a swearathon outside The Lowry to highlight its absurdity.[23]

The Glossop Brook is constant background music as we stand in the space with many names. After sharing hopes for the future of Glossop we walked back onto the dreaded main road past the Vivienne Westwood mural. She was born in nearby Tintwhistle and the artwork became a shrine after her death. Locals were really proud of her, and chuckled at how the area wouldn't be imagined as a hotbed of punk. Should there be a museum? I was grateful the trains were running, as more than once I've relied on a lift or the painfully slow rail replacement bus.

I share the same wish for Glossop as I do for Ashton and Stockport: that they learn from their neighbours in central Manchester and aim for development that benefits all the residents. I hope they embrace plurality and leave room for dreams and visions. I hope decision makers can have an open mind and take the time to listen, really listen, to what local communities, and the aural landscape, is telling them.

5

Sheffield

Nature Walk: Wherever you are, look for signals created by the non-human. Desire lines left by snails, birds nesting in concrete, cosmic maps made by lichen, the mutuality of mushrooms. If in doubt stop and admire a pigeon then try to follow where it leads.

Sheffield is a hug of a city that smells of Hendersons and petrichor. I lived in Leeds for much of the 1990s and it was always in my peripheral vision. It was a toss of a coin that took me to Manchester instead. Actually, that's not quite true. It was a combination of cheap housing and a need to escape bad choices I made in Yorkshire that swung the decision, but Sheffield always nestled somewhere in me. I thought I would probably live there one day, surrounded by hills and nurtured by creative collaborations, but that never quite happened. Instead, I deepened our relationship by studying for a PhD there. Prior to starting my studies I'd found myself with a different set of choices about which direction to go, and so one Sunday I stood next to a hexagonal concrete building on the edge of Weston Park. It sang to my heart somehow and I knew this was where I needed to be. I'd come to have a desk in that building, in a room full of desks, each one piled high with books, plastered in Post-it notes and personalised with mementoes. As I typed I could hear the ducks in the park; when their quacks annoyed me it was time to turn away from my laptop and go for a walk.

Sheffield is often lauded as the greenest city in the UK and walking infrastructure there sometimes feels more thoughtful than in other cities. As I get off the train I smile to see a giant Hendersons Relish bottle welcoming weary travellers, breaching my dislike of billboards. I love Hendoes, it's a delicious condiment and helpful plant-based substitute for the umami of Worcestershire sauce. It embodies the local terroir and used to feel a bit secret. I would always stock up on a visit, but these days you can find it in supermarkets everywhere. Their factory used to be on the edge of the university campus and I recall Victoria Henshaw[1] talking about the smell. Victoria's work focused on scent, how it shapes our perceptions and reveals our preoccupations. Contemporary cities often smell sterile, homogeneous, apart from blasts of baking (often artificial) or cosmetic shops. One of my first jobs involved decanting perfume; it played strange tricks on my nose and stripped the paint from my gold DM boots when I spilt it. It was Victoria who made me think about smell differently and inspired me to study at Sheffield. Her smell walks, and smell maps, can sniff out differences, the politics of place are carried in the air. Scent is unruly, leaky, uncontainable. It wafts over barriers. Stop and sniff: Who is cooking and eating here? Do those flowers have a scent? What does the aroma of that artificial air freshener convey? Bland? Sickly? Clean? Exotic? Who is that perfume attempting to entice? How does the river smell? Would you drink it? Why are water companies failing us and leaving us to swim in shit?

Coming out of the main railway station I smell Springtime and coffee on the wide empty plaza. A curved metal structure borders the path. The mirror reflects the industrial past of a city linked to steel, a knife and spoon smoothed by the years, sinuous and streaming with water. Cross near The Showroom cinema and head up Howard Street. Billboards proclaim CGI previews of 'The Future Space' where phantasms walk buildings that are yet to exist. Along the edge of Sheffield Hallam University campus and over the road to the Millennium Galleries. Handy, and accessible, public toilets and then up the escalator, through the gift shop (of course) and out into the Winter Gardens.

A cornucopia of green, tranquil yet buzzing with life. There's a generous smattering of benches and a group of around a dozen school kids are squashed together on and around one of them. I wonder if this is an officially sanctioned trip or a skive. Elsewhere, individuals and couples eat packed lunches, make phone calls, speak in whispers or laugh uproariously. Last time I visited there was a special snooker event happening and I am glad to see its rope fences and television cameras are gone. There is no ambiguity: this is indoor quasi-public space but I can't resist its charms. It's full of people just being, dwelling, the gardens are part of their day and they feel welcome to claim their place. There's no entrance fee and the automatic doors glide open as you approach. The bird of paradise plant lured me in for a closer look at their astonishing plumage. It looks, feels, like a diverse crowd, not really paying much attention to each other. The atmosphere is light, airy, expansive, despite the greenhouses temperate state. If I was a purist I would snark: this is not an unprescribed, spontaneous gathering place. Footpaths are clearly marked and security, however genial and low key, is very much evident. The Winter Gardens may not be a textbook public space – it might be glass, but it still has a roof – however, it's clearly loved and offers a vibrant sanctuary and nourishment to the city.

Out again, past the Peace Gardens. The water features, fountains and cascades, seem less erratic than those in Piccadilly Gardens. Embedded in a wall and on the ground are standard measures: pole, perch, links, feet, designed to settle disputes about the length of goods. Onto Barkers Pool, past the City Hall and The Women of Steel Statue. Up Devonshire

Street towards the green and I feel a pang passing Rare and Racy book and record shop, a vacated treasure trove lost to redevelopment plans. As I cross onto West Street I wonder why I didn't take the tram. It's been a while and my memory had smoothed the hill and shrunk the distance. The tram didn't occur to me because I miss Sheffield and feet on the ground enable me to reacquaint myself in a far more satisfying way. During the pandemic, Sheffield-based Longbarrow Press[2] offered a home delivery service of their exquisite books. Founder Brian Lewis documented his walks across South Yorkshire. Elsewhere in the city Joanne Lee[3] documented the everyday experiences of her lockdown. Combined, their online diaries gave me a delicious glimpse of the place, but it was no substitute for being here.

One of my favourite works Longbarrow have published is Pete Green's *Sheffield Almanac*[4] in which similar questions to those I often focus on are explored: what makes a place distinctive, gives it a peculiar character, the interplay of people and landscape and the impact of industrial decline, change and regeneration.

> The rivers, forever five rivers
> > Driving down through limestone, carving grit –
> Every vista the city delivers
> > Forged at their silvering hand
> And while kids strike out in creative
> > Trades their grandads, roiling over molten vats,
> Wouldn't't've bloody dreamed of, the landscape's native
> > Relief, scribed into five valleys, is the constant.

* * *

Sheffield proclaims itself to be a 'City of Makers'. Like everywhere else I have visited, it has changed drastically over the last few decades as traditional manufacturing and industrial jobs declined. Pete's quote above speaks of the gulf between generations, the changing definition of work. Sections of the tram line feel like a border between those benefiting from a service economy and those who work for it. I don't remember The Hole in The Road but it recurs in conversations about what has been lost in the city. It was another 1960s concrete utopia that was never fully realised. A subterranean convergence of several underpasses, a large circular area with no roof – open to the elements, the hole in the road. One of the walls housed a giant aquarium, which is recalled fondly by those that saw it as a child. However, in a way familiar to reminiscences about Piccadilly Gardens of the same era, people often caveat their stories with complaints. It could be disorientating down there, had a reputation for nighttime violence, unsavoury characters gathered on the benches. The nostalgic glow is tempered, and I am sorry I never visited to form my own opinion. What was once The Hole is now buried under Castlegate Tram Stop.

Another Sheffield icon of modernist architecture still stands today. The Park Hill flats are emblematic of gentrification processes across the UK. Built as social housing, and home to families who made lives and communities there, they became stigmatised and ill maintained due to Thatcher's housing policy. Today, Urban Splash have renovated part of the estate, with the rest still awaiting its turn. Graffiti has been shamelessly replaced by neon.[5] Park Hill has always been a landmark that told me I was in Sheffield, a reassuring presence when I was a stranger unable to navigate across the city. I remember spending what felt like hours searching for The Boardwalk. I swear it seemed to move between my visits. It's gone now, nothing but an echo of the music and good times left. In Vancouver, Kristina Rothstein goes *In Search of Lost Venues*,[6] walking neighbourhoods with musicians who played in places that have now closed. An enthusiastic music fan, Kristina does this because 'the specific energy of certain venues which inspired music scenes has been transformative in my life. The right music in the right venue can create ecstatic experiences.' I feel what she means acutely. The club as crucible, music as alchemy.

Once I was with a friend who drove us to Sheffield from Leeds. We were going to see The Handsome Family. A magnificent band who manage to sound utterly timeless, both contemporary and ancient, magical realist and searingly brutal. We got so lost en route to The Boardwalk we were pulled over by a police car. They escorted us to the venue with a warning that if they started their siren or flashing lights we were to pull over immediately and await further help. What strikes me now about that memory is less the jovial policemen than our naivety. It wouldn't be many months later that I would have felt fear at being noticed by the cops. It's a luxury only experienced by a few, almost always, white, folk.

That illusion was shattered the day I saw a police van clip the back wheel of a bike being ridden by a teenager on my way home in Whalley Range, Manchester.[7] There had been some shouts about a stolen bag from people leaving a nearby pub and the van came whizzing round the corner. It swerved to hit the bike, and when the lad tried to get up they grabbed him and slammed him into the side of their van. They were not gentle and polite. It's such a cliché I'm embarrassed to spell it out but they were white, their victim was Black, and they barked at anyone who stopped to watch. Ridiculously, I went home, unsettled and unhappy, and after some contemplation I rang the police to complain about the police. The voice at the end of the phone mocked me for not having a registration number and warned there would be consequences if I made a formal complaint, which would be pointless without that evidence anyway. Later, I would witness first hand police brutality at protests and disproportionate surveillance of protesters.

The SpyCops (undercover policing) inquiry is ongoing as I write. Sheffield knows these horrors through Hillsborough and Orgreave, two of the largest scale miscarriages of justice in living memory. My experiences are trivial in comparison but represent to me a clear break in my understanding of how our society works. Corruption and violence need challenging wherever they are, whatever the scale. We can care about safety and still want to defund police, prioritise social services, support networks and mutual aid. We also need to treat violence against women and girls more seriously, but changing the law or increasing punishments without addressing underlying issues won't be enough. These things can all be true at once.

I'm snapped back to now by the sound of the Sheffield tram. I love the names of the stops: there is poetry in Halfway and Crystal Peaks, especially when said in a Sheffield accent. I always seem to end up in a conversation, free with every ticket. I resist jumping on and head over instead to the Arts Tower. It's a gorgeous bit of modernism, elegantly proportioned and shining in the sun, and it houses one of the last working funicular elevators. It's a local landmark because it is so rare, as Kiera Chapman[8] told me, Sheffield is low rise, a 'town like in scale ... you never feel very far from the Peak District'. This is in stark contrast to Manchester's vertical expansions. Sheffield feels more intimately connected to the surrounding landscape. Kiera tells me the local authority has done a great marketing job, creating the outdoor city as a solution to post-industrial malaise. They emphasise Sheffield as the perfect base to go rock climbing, hiking and appreciating the joys of the outdoors, which, as Kiera tells me, can often feel quite ableist. It also chimes with a romantic notion of the sublime walker which we both dislike immensely. Like the flâneur, this identity is remote and unrelatable for many women. Kiera puts it like this:

With walking where I get frustrated is this kind of emphasis on a sort of Bronte-isation, one should just be free in the environment … it's slightly puritanical … one must do it for one's health … I've never ever been for a walk on my own as a woman and not been aware, at some level, of where people are around me … I can't just run free in a floaty dress over the moors screaming like Kate Bush, amazing though she is … I have to be aware, I have to be conscious of whose around me.

Kiera is absolutely not disconnected from nature. She is a co-author of *Nature's Calendar: The British Year in 72 Seasons.*[9] It provides an alternative vision of time, seasons lived by, with and through nature. The diary of 'micro-seasons' shows what happens in minute detail and tells the stories of flora and fauna in the UK. The book emerged from a Twitter project, and as X becomes increasingly problematic we lament the demise of the platform. Another venue closing, but the conversation will continue to flourish elsewhere. Kiera notes nature is every-where, entwined in our lives. She says paying close attention to the urban environment can be massively illuminating for botanists of any level because 'some of the really interesting stuff grows in the dog weed zone or in crevices in walls.' Kiera tells me she has spotted 'incredible stuff' thriving in suburban streets and on disused railway sidings, and we share the glee provoked by these encounters with unruly outbursts of nature.

Kiera's point about a puritanical edge to walking for health or accessing the countryside is an important one. There are undoubtably huge wellbeing benefits – physical and mental – to walking and accessing green space, but it is not a mythical cure. There is no sudden epiphany triggered just by setting foot outside. Indeed, for many disabled and chronically ill people, walking can be unwelcome and at worst harmful. WalkCreate survey respondents told us: 'Doctors and others are always telling me to go for daily walks and I find it frustrating and upsetting, because all that trying to walk every day did was cause me more chronic pain and make me bored and stressed' and

> There is a lack of social awareness around the fact that for people with chronic illness some days a simple walk to the shop is too painful. The cultural and societal push for 'counting steps', 'logging miles walked' etc. all in the name of health and wellbeing can be detrimental to the mental health of people with chronic illnesses ... Everything feels like a competition and like you are being silently judged.

The notion that simply being there in nature activates some kind of epiphany is grossly misplaced and can be actively harmful. Green does not always automatically equal magic. It's hard to get a sublime revelation if your back hurts, you have a hole in your sock or a swarm of midges attacking.

There can undoubtedly be positives for many of us from both a connection to nature and walking. However, wellbeing narratives are too often individualistic and ignore structural issues. They can do disabled and chronically ill people real harm. I frequently walk past a poster for a gym that says, 'Your Health is Your Wealth, Spend it Wisely' and I groan at the message this sends out about which bodies are valued. Recently, I have been working on a project led by Bethan Evans.[10] It looks at the experiences of women with energy limiting conditions (ELCs) and we have heard horrific stories of the harm enforced exercise can do to people with ME or similar conditions. We've also been thinking through crip time, the idea that for disabled and chronically ill people conventional time structures are fragmented and distorted. Some tasks simply take longer due to how our bodies work. I've lost friends when they have had no patience for my pace. It's humiliating, hurtful, to be five minutes behind and perpetually struggling to catch up because they will not slow down. Any pause for me to reach them was never leading to an actual rest for my benefit, just a momentary exasperated sigh from them before the next stretch. However, crip time is not just about slowing down, as Ellen Samuels says, '*Crip time is time travel. Disability and illness have the power to extract us from linear, progressive time with its normative life stages and cast us into a wormhole of backward and forward acceleration, jerky stops and starts, tedious intervals and abrupt endings.*'[11] Conditions fluctuate, bodies age, crip time demands new rhythms. My personal experience is that crip time also includes huge amounts of waiting: for appointments, for referrals, diagnoses, services, to be seen when actually at a clinic. This can be painful, but for me it is mostly very boring and frustrating.

There is never one essential or standard body, and as ever care must be taken to avoid generalised assumptions. Walking can, and does, provide respite, joy and sustenance for disabled people. Steve Graby[12] works for a disabled peoples' organisation and writes from their perspective as a non-binary autistic person. They share how important walking is to them, and how it can be almost a compulsion. They often feel happiest, and most fulfilled, on a long hike. However, Steve also cautions us. For neurodivergent people walking can be medicalised and viewed as a both a symptom or diagnostic tool which pathologises an individual's autonomous choice of movement. Wandering while autistic is constructed as problematic in a way it is not for others, and so they may find themselves restricted or censured by misguided and damaging medical assumptions.

Louise Kenward is a visual artist, psychologist and writer based near the coast of Southern England. Much of her current work is concerned with navigating walking while living with chronic fatigue and pain. These are landscapes she loves, and places she feels an intense connection to. Her writing captures the coast, and people, in flux. The ebb and flow of the sea, deep time connections to processes of eroding and dispersal. In *100 Tiny Oceans*, she walked to the edge of the sea 100 times and captured water from a wave in a jar. She documents incremental differences and nurtures a closer understanding of marine life. Rockpools, seaweed, flotsam and jetsam. This is an extract from *Tiny Ocean #43*.[13]

> The beach: a place that hardly exists at all, washed away twice a day it is largely hidden from sight. My fatigue: a condition that is barely there, unseen, it is hard to believe in, but it lingers and dwells beneath the surface, rooted firm. As real as the ocean, it is as hard to hold on to.
>
> I begin to be able to walk to the beach, it is a marker of achievement. Being by the sea is a joy I can no longer take for granted. I stand at the edge and take a photograph out to the horizon. It's evidence.

Louise edited *Moving Mountains: Writing Nature through Illness and Disability*,[14] a profound and important collection. I could fill these pages sharing the wisdom but instead I wholeheartedly recommend seeking it out. Many of the chapters share experiences of walking, or not walking. I was struck by Isobel Anderson in *Not Healthy, Never Healed*, walking with the River Ouse and saying,

> I could momentarily opt out of the usual human performances of both gender and sexuality. The loss or change in both of these identities were one of my most fundamental experiences at this time ... To walk for hours and not be seen by others helped me to, in some small way, accept the loss I felt.

Her walks were 'a modest, but crucial, form of self-expression at a time when I felt shut out of my own life'.

Khairani Barokka (Okka) shares powerful testimony in *The Clocktower and the Canopy*. She begins relating her time in a London hospital where she explains: 'What built up Big Ben broke down my bodymind. And it was through rainforest trees.' She speaks of the harms colonial violence did when it enslaved her ancestors from the Indonesian Archipelago. It disabled people, destroyed the environment and shattered connections where nature and culture are entwined. Okka speaks of generational trauma, collective memory and resistance to the ravages of extractivism and exploitation. We need to listen, and learn, from cosmologies that understand the human as not apart from, but a part of, the natural world.

Okka's words are a powerful reminder that the hardest barriers to dismantle are not always physical. One of the organisations we worked with during WalkCreate was SEM (Sheffield Environmental Movement). SEM's mission is to 'work with Black, Asian, Minority Ethnic and Refugee (BAMER) communities and environmental organisations to ensure everyone has a clean, healthy environment and access to open green spaces'.[15] Founder Maxwell Ayamba told us: 'there has been a lot of Black presence in Britain for centuries ... we're walking to reclaim the land our ancestors have walked for centuries, but yet have not been written into that landscape ... where Green becomes White, then there can never be diversity'.[16] Maxwell is part of a movement to ensure countryside in the UK is welcoming and accessible for all.

Rhiane Fatinikun founded Black Girls Hike UK after travelling through the Peak District on a train and vowing to explore the landscape. She wanted to walk with a community where there was no work needed to fit in and where she would not experience micro-aggressions or be expected to represent, because 'many of us spend our days in jobs where we code switch to fit into a majority white environment so, when it comes to our leisure time we don't want this extra layer of labour – we just want to relax'.[17] *Finding Your Feet* explores her personal journey, and those of other Black Girls Hike members. The joy and solidarity radiates from every page, but the book also offers a very practical 'How to' guide. Rhiane generously shares skills such as navigating, map reading and what kit is needed. Black Girls Hike recognise absence and inequality at all levels and offer training for outdoor leaders. Their hugely popular programme of activities includes hikes, camping weekends, a wide range of other outdoor pursuits and international expeditions.

The whiteness of the rural landscape links back to romantic histories which obscure more complex and exploitative trajectories. They are challenged and restored by groups like Black Girls Hike and their kindreds Muslim Hikers, SEM and The Wanderlust Women. The latter was formed by Amira Patel who said, 'I had never come across another woman who wore a niqab or hijab to go hiking, … It made me realise that a lot of women aren't going outdoors because they don't feel confident, or they don't think it's for them because they don't see anyone else who looks like them doing it.'[18] Amira designed the first hijabs and niqabs specifically for trekking conditions and The Wonderlust Women host journeys across the world as well as within the UK.

Artists such as Ingrid Pollard make visible absences and colonial legacies. Her photographs powerfully deconstruct the stereotypical assumptions of pastoral whiteness and isolationism. The caption beneath *Pastoral Interlude No 1* (1988), an image of a lone Black woman sat within the supposed rural idyll, says, 'it's as if the Black experience is only lived within an urban environment: I thought I liked the Lake District where I wandered lonely as a Black face in a sea of white. A visit to the countryside is always accompanied by a feeling of unease, dread.'[19] *Pastoral Interlude No 5* (1988) depicts a figure wading in a stream, fishing, with the words 'searching for seashells; waves lap my wellington boots, carrying lost souls of brothers and sisters released over the ship side'.[20] The horrors of the transatlantic slave trade linger and continue to shape the English countryside. The walks Corinne Fowler takes in *Our Island Stories* map those violent histories and their lasting legacy on both identity and landscape across the UK.[21] The threats and abuse she received for research on the National Trust's reassessment of its colonial links demonstrate how far we still have to go.

* * *

The question of who belongs here can also be viewed in terms of legal ownership and access rights. On the train over I passed Kinder Scout, site of one of the most successful acts of civil disobedience in UK history. The Kinder Trespass in 1932 was hugely significant, but is often depoliticised.[22] Benny Rothman and the others were communists and this was part of a wider liberatory struggle. After their famous wander on Kinder they went back to fighting fascists. Rothmans' Jewish identity, and how this may have contributed to his treatment by the authorities, is also often downplayed. There were other important protests in the North West too, including Winter Hill and the Elton Reservoir. However, Kinder is rightly celebrated and has become a totemic victory on a long journey towards the right to roam. An annual event, *Spirit of Kinder* combines commemoration, celebration and a contemporary call for action. In 2024 (Kinder 92), *The Kinder Pledge* was launched. The text was agreed by a wide range of campaigners and campaign organisations concerned with outdoor pursuits and access issues.

We declare our commitment to a Universal Right of Free and Responsible Access to all landscapes, Rural and Urban, for all.

We pledge to peacefully campaign to enshrine in law the right of universal open access to land and water, urban and rural, similar to that in Scotland and other countries. We will also support the actions of those who undertake forms of non-violent direct action to build support for such a law.

We pledge to do no damage to any landscape and to recognise that some areas may not be accessible to everyone at all times, in order to protect the environment and wildlife.

Through formal and informal programmes of education, we will seek to extend popular understanding of the natural and built environments and how we can best protect and enhance the rich diversity of the life and lifestyles that they enable and support.

Individually and with others we will explore, enjoy, exercise and simply be in open urban and rural space. We will campaign to ensure that open spaces are welcoming and accessible to everyone.

I am a proud signatory to the pledge and attended some of the meetings it grew from, facilitated like the event by the Heyfield Kinder Trespass Group. It emerged because of frustration that land access and a right to roam was not on the political agenda. Only eight per cent of England is covered by the current legislation, and much of that is in remote areas that are tricky or impossible to access.[23] Right to Roam campaign nationally and locally to improve this shameful figure. Crucially, they want a right not just to walk in the countryside but to swim, loiter, play and camp – a right that they fight passionately for. I spoke to Nadia Shaikh, part of the national Right to Roam team. She talked to me about why this matters so much and what she feels the benefit would be to a more equal land system.

It's mind-blowingly complex but it's also beautifully simple ... when you can be rooted in a place and genuinely have a sense of belonging and safety, humans will naturally care for the environment around them ... given agency and access the natural leaning of a community will be to care for the shared resource of the people and the land around them ... It's the systems that corrupt that. It's within all of us and not beyond reach ... at Right to Roam we are trying to reinstate fundamental rights to do a whole myriad of activities and from there we can build infrastructure genuinely based on how people need to be outside.

This is bigger than merely a footpath campaign and it is not just about walking. This has led to some conflicts within groups who have slightly different agendas, but work such as *The Kinder Pledge* shows there is much common cause and solidarity. I have personally felt frustrated that I have had to spend so much time and energy fighting micro-enclosures or to preserve specific rights of way such as at Ralli Quays (see the concluding chapter). Sometimes I worry this deflects from the bigger point of principle, but I do believe at the moment hyper-local fights are necessary. If we lose a space or path because it is extinguished to allow building or is annexed by an individual it is harder to win it back. These are the places we personally use, care about and connect to, but we do need to remember they are also part of a bigger picture. Nadia says, 'As activists we become semi-historians' as we learn about what matters to us, and about the history of wider struggles. However, we can't always know how our own work contributes to a movement. She says at the time our predecessors were fighting for 'civil rights, women's rights, those people didn't see themselves as a movement, they wanted one thing. Lots of us are asking for things at the moment, history will tell the tale of what we're doing'.

Nadia sees synergies between groups fighting the climate crisis and improving biodiversity, Right to Roam themselves encourage *Wild Service*.[24] This is about 'rebuilding a culture of care and connection with the natural world' and they encourage people to act as guardians and work to support and restore nature. Ultimately, Nadia suggests this might be viewed in the future as an anti-capitalist movement because 'It's all about stopping extraction and exploitation which is the fuel of capitalism.' To me this a fundamental link back to those foundational ideas of the SI and psychogeography, to the need to remap our environment carefully, collectively and without financialisation at its core. The habits of capitalist domination can be hard to unlearn. Nadia talks about the need to 'decolonise your mind' and be open to ideas of communality and co-operation. Nadia felt the psychic impact of cultural norms herself after moving to Scotland. It took her a while to decompress and acclimatise to not being restrained by the repressive land laws in England.

Very often psychogeography is viewed as an urban pursuit. This is challenged by artist Jane Samuels.[25] Jane shares her walks on Instagram and the visualisations of rural places in her *Terrain: Anatomical Landscapes* are stunning. She says, 'I understand the premise, the idea psychogeography is primarily a urban conceit ... but the idea that the fundamental politics and practice don't apply beyond the urban is for me a nonsense.' Jane makes a very important point about the dangers of a false binary, and how questions of access and equality resonate throughout the UK.

I would argue most, if not all, the concerns you are dealing with in urban space around private and public already existed in the rural long before the urban happened. [In my work] I'm thinking about place in terms of who owns the land? who is allowed on the land? who belongs? who doesn't? how do people become empowered or disenfranchised in place? who belongs where? how does disability fit in on the fell? – the answer unfortunately is sometimes it doesn't – and how do people of colour feel in the rural environment? The same conversations we've been having about urban space for a long time. It probably came from the country first.

The symbiotic nature of the rural and urban is highlighted when Jane talks about how water from the Lake District is transported to supply Manchester with this most vital resource. Jane's definition of psychogeography also resonates very deeply for me as a place where:

Our consideration of place and space interacts with a political understanding of our environment and the power structures inherent (in them), how all those things interact together ... For me it offers almost a responsibility to push back against those power structures and it's an opportunity to play in place where we might not otherwise do so. This is disruptive in and of itself in places where we might not have power.

Jane also told me: 'There is something radical and wonderful about the idea of often female and often disabled radical walking family who are supportive and interested in what each other do.' Her final point is important because it speaks to something this book addresses head on, which is that the work of many psychogeographers and walking artists has not been widely recognised. Too often accounts rely on star names rather than lifting up a more diverse, and I think more interesting and valuable, range of work. I've felt the support and encouragement Jane speaks of and I strive to offer that helping hand to others. I remember attending a *Walking Women* event in Edinburgh, and feeling nourished by the mutual respect and shared understanding. There was no need to justify why walking or why women. However, regardless of critical acclaim and conventional notions of success I have no doubt critical walking art will still be made and mutual support given. Like the flowers that force their way up between cracks in the pavement we will keep creating and sharing our stories.

For Jane, her work is a deeply political act, and she extends her compassion and liberatory ideas for non-humans too. She offers a powerful perspective on interspecies relationships when she says,

> My politics is also a history of animal rights and veganism and so it's the same arguments but the [rural] context differs to the urban because it becomes more visceral, the animals are there in place ... how can we widen our focus to consider the place of non-human animals in their homes and environments and what we are doing to those environments.

Jane's work often asks tough questions about the morality of rural tradition and recognises the incredibly strong tensions around activities such as fox hunting. She believes firmly that 'preconceptions about who has power in rural settings should also include non-human folks'.

If, as Jane suggests, we need to consider the more-than-human we also need to consider what we view as natural and nature. This book focuses on the UK where humanity has impacted the landscape for centuries and the truly wild is vanishingly rare. We need to think about the quality of outdoor space – and how the countryside relates to the urban. There is a symbiosis more complex than we often realise. Kiera

Chapman brings me back to Sheffield and the Peak District. There is huge unease as the land has been massively degraded by human activity. She points directly to grouse moors, heather burning, and land ownership. The practice of keeping grouse impacts everyone in the City of Sheffield, contributing to flooding and pollution, creating 'such dead places' on the moors. We may consider the rural landscape here as sublime, but we don't usually see the practices of land management or recognise the absences, the lack of biodiversity. Kiera wants action on social and environmental inequality, which she firmly posits is an inevitable consequence of the current land ownership and management systems. The rural and urban have a reciprocal relationship and are entwined, complicit, in so many ways. The boundaries are blurred when we think of factory farming, holiday homes and online working. Rates of poverty, isolation and deprivation are devastatingly high in many areas of the countryside. There is no rural idyll, there never was.

The need for what Right to Roam call *Wild Service* should not be lost in more philosophical debates about what is 'natural'. Wherever we are, the more-than-human world touches us. In Sheffield, there was anger about the removal of street trees, and similar arguments play out in other places too. Jo Norcup[26] has been documenting the trees of Beeston where she lives because 'to be in the company of an old tree is something to be revered'. She has years of experience in environmental campaigning and feels if people know individual trees and name them they feel more responsible, more connected and able to fight for their preservation.

In the *Meadow Behind Bars*, Alison Lloyd[27] documents the walks she took during lockdown in a meanwhile space or accidental meadow. This was a patch of urban wasteland that had been neglected while awaiting redevelopment, giving flora and fauna a chance to flourish. What had previously been nothing more than a shortcut became 'my friend, a place of solace and recuperation as I worried about family and money'. I think about that scrap of land in Shudehill I shared in Chapter 1 and how easily it can be overlooked. We are lucky it is still there (for now). In Deptford, Anita Strasser[28] documented the impact of austerity and gentrification on her community. This included powerful testimonies about the occupation and eventual eviction of a community garden. I never visited but could feel the rage and pain of its loss.

Back in Manchester, nature is entangled in the processes of gentrification. I watched the Deansgate Square Towers slowly rise up, transforming the skyline and blocking the wider vista. Horizon shrinking. Part of the development opened up access to the River Medlock, as did a new unrelated student tower further down in Circle Square and the Mayfield Park I visit in the concluding chapter. At Circle Square, where the BBC used to be based, the route has a lockable gate, when it closes at night the illusion of equity is broken. Deansgate Square is different, no gates, or visible border, but as you approach the texture of the pavement changes and the glass towers loom. It bears all the currently fashionable hallmarks of a homogenised privately owned public space (POPS). Walking there with a friend, architect Lorenza Casini, I laughed at the absurdity of a section of plastic grass for dog walking.[29] My laughter rang hollow as she told me about the environmental impact of artificial grass. A young woman resident saw us inspecting the ornamental (and probably ineffectual) wind breaks. With an enthusiastic smile, she told us how much she loved living in the towers, but had to lie about her income to be allowed to sign a tenancy. She was excited by all the amenities: an in-house concierge, laundry service, gym, shops, bars and living rooms to rent if you want to invite folk over for tea because there is no room in your apartment for guests. We didn't want to dull her effervescence but afterwards felt uneasy about the hermetically sealed luxury ecosystem and its relationship to the rest of the city.

Further south, in the suburb of Chorlton, there is an ongoing campaign to save Ryebank Fields from development by Manchester Metropolitan University (MMU). The LRM have First Sunday'ed there in a joint wander with our local Right to Roam group. It was one of the biggest crowds and most remarkable places we have visited so far. I spoke to Sarah Benjamins from the campaign who shared why this particular place matters. She told me: 'the thing that the field does, one of the amazing things about wild nature is that it looks after itself and it's kind of magical ... this place thirty years ago was just a football pitch and now it's full of abundance ... a fully functioning ecosystem'.[30]

Sarah felt that the fact that Ryebank Fields is in the city means this is both a social and spatial justice issue, it's providing access to nature for people without cars or the need for a long journey. Her description of Ryebank Fields reminds me of the arguments for other public spaces: 'it gets under your skin ... Nature connection matters in the city, one [reason]) is around community ... it's a public space, I think nature places in the city are really good third places, always there, with the opportunity for low pressure interactions with others.' Sarah talks about a shared sense of awe and a moment of connection between strangers if they both spot a sparrowhawk; it's 'the shared sense of I love this place, and you love this place ... even if we never speak that's our connection'.

Sarah is very concerned about the notion of 'genuine' nature spaces in the countryside or green belt being viewed as the only ones that need statutory protection. I agree that this undervalues the immense value of many urban and suburban green and blue spaces. She dismisses the notion that saving Ryebank Fields comes at the cost of housing opportunities, as there are many nearby brownfield sites and empty properties. She is proud to be 'unashamedly defending nature' and believes 'connecting to land is resistance because the way they shut us up is ridicule or persecution'. She demonstrates the harm micro-enclosures cause and why we need to keep fighting them. This campaign is local to me, but there are myriad others across the UK.

*   *   *

I think community actions like the Ryebank Fields campaign matter precisely because they are how we connect with where we are on the planet and who our neighbours are. The challenges of climate chaos and a burning world are being tackled by women walking artists and activists in a number of ways. I'm still thinking about what the countryside means as I head back to the station in Sheffield. As I wait for my inevitably delayed journey home, I think about some of the wider barriers to accessing the countryside. Getting there. Public transport is often very poor, but parking is even worse and cars dominate the landscape. Where actually are we trying to go anyway? How rural is a village full of remote workers or holiday lets, and what about industrial agricultural factories?

The rural idyll is just a nonsense but somehow it has become part of our national identity. The short version is I blame the Romantic poets, and capitalism – the invention of work needed a counterbalance and whitewashed sugary nostalgia has always sold – and this became reified over time. The not very distant echoes of colonialist practices, feudal systems and patronage feed into this too. Spoiler alert: there is no golden agrarian utopia to get back to. It's saturated in violence. It feels like so many of the countryside myths come back to Gemeinschaft and Gesellschaft and the foundational idea of cities as dangerous, corrupting and intrinsically evil. It's a paradoxical dream (posh word for scam). If we all moved to the countryside it would be destroyed. We are an urban country, on an urban planet. If we concentrated on the beauty of where we actually are, and how to improve here and now instead of plotting an escape to the country, imagine what could be possible? We also need to think about which flora and fauna we value and why. How long must a species flourish until it is no longer branded non-native? I remember a walk in Sheffield with Sheffield Psychogeography Action where I learnt about fig trees by the River Don and community eco-gardens.

As the train finally leaves Sheffield, I think about Anita Sethi's[31] walk through the Pennines, 'The Backbone of England'. In *I Belong Here*, she documents a walk motivated by racial abuse she was subjected to on a train like this one. Her walk was a reclamation, a defiance, a way to walk herself into a place where she always belonged but had been alienated. Anita documents 'how a landscape plants itself within us, its roots growing within the heart, its branches arching through the mind, its rivers running like blue-black ink through the veins, lifeblood'.

I think too about Doreen Massey reminding us: 'Feeling you belong to a place in no way entails that it belongs to you … Ask not "do you belong to this landscape?" but "does this landscape belong to you?"'[32] The systems that underpin enclosure, imperialism and colonialism are patriarchal. As Nadia from Right to Roam says, 'it's one of supremacy – an able bodied white cis man being dominant in the landscape – you can imagine the roll call of people under that'. Exclusion and extractivism are written in the fields and hedgerows just as they are factory chimneys. We need to be there, in the fells and fields and woods, to begin rewriting a new countryside code.

My walking prompt for this chapter was to focus on, follow and learn from the non-human. They are not as separate, or subservient, as we may assume. When I'm feeling unsure of myself in the city, I often seek out buddleia, the patron plant of

The LRM and loiterers everywhere. Endemic to Asia, Africa and the Americas, they have self-seeded from gardens. We have an annual no-prize competition to find he most audacious 'butterfly bush' as they thrive amidst dereliction and crumbling masonry. Marie Pattison was half of the Manchester Zedders who explored the city map square by random map square and shared their adventures via a blog.[33] We have become friends and I remember her telling me that when she first moved to the city, she looked out of her bedroom window, and saw buddleia growing on a rooftop; she thought if it can thrive here, so can I.

6

Liverpool

Ask someone to direct you to the heart of the landscape. Get there without using a map or your phone (unless someone draws you a map, in which case that is brilliant luck).[1] When you have found the heart try to discover the rest of its body and how it extends outwards to connect with others.

I don't remember the first time I ever knew the world wasn't built for me but it's something that I realise anew every day when I encounter steep steps, broken lifts or roads with no crossing points. I'd had to rely on a random stranger to help me off the train at Newcastle and was still thinking about this when I got to Lime Street. Just before the station you go through the Edge Hill Cutting, a sandstone gorge shaped, strengthened and reinforced by generations of toil. Deep time glimpsed from the train window. The stone is strong and true and I get the same strange body glimmer I feel on beaches or cliffs. A connection resonates through time and space. Across the city are the Calderstones, older than Stonehenge, now in a park, decontextualised but oddly affecting just standing there. Along the coast are beaches where pebbles mingle with debris from the Liverpool Blitz, the tides slowly eroding distinguishing marks. As we pull in, I notice the station clock is broken, ironic as the railways were one of the key reasons we moved to GMT and standardised times. Before 'train time' was introduced, timetables were confusing – solar time utilised sun dials and was necessarily localised because of geographical differences in celestial movements.

Lime Street Station concourse always feels a bit underwhelming for such a vibrant city, it needs more of a fanfare. Oh Liverpool. Maybe the locus of even more clichés than Manchester, but many of them feel true. In my experience, it really is an incredibly friendly and politically conscious place. Many have held fast with their boycott of the *Sun*. The boycott was called because of tabloid lies about the Hillsborough disaster and we could all learn from it. Debates about media bias are not new, and have accelerated with the advent of social media. The spectacles' tentacles infiltrate everything, layers and layers and layers before we find the truth online, in print or broadcast. Near the station entrance is a statue. Two travellers frozen awkwardly together, it's unclear if they are in communion or coincidental transit. The man is immediately recognisable to me, he's comedian Ken Dodd wielding his tickling stick. The woman is less familiar – she's middle aged, smart, serious looking, holding a handbag and an egg. Not the kind of person you tend to see in bronze. She reminds me of an old school *Coronation Street* matriarch. Bessie Braddock was an MP and a fierce campaigner on housing, social and health rights for the people of Liverpool. Censured for calling a Conservative colleague a liar, and known for her campaign tactics such as bringing a megaphone to debates, she said, 'if you didn't do something outrageous, nobody would take any notice of you'.[2] However, she also championed the drowning of Capel Celyn, one of the last Welsh-speaking villages in North Wales. It was sacrificed, despite objections, to provide water for Liverpool and the Wirral.

Some of the York stone paving slabs on the concourse outside the station have carvings on them. One is of a CCTV camera, appearing to me like an alien on a tripod. Opposite is St Georges Plateau, outside St Georges Hall. The pavement here is shot through with rivets and furrows, evidence of prehistoric waterflow, and there are fossils visible in the limestone bollards. It's long been a gathering spot for celebration, commemoration, proclamations and protests from Suffragette meetings to the huge general strike of 1911 and more recently Extinction Rebellion and student climate strikes. Today, however, it's covered by the Christmas market which I won't be visiting, not least because I need to get to work.

When I'm going to work, especially in the morning, my walk isn't creative or revelatory. It is a brisk, functional A to B, transporting me to my office. I'm more alert to traffic than hidden connections and ambiences. That alertness isn't always enough. A recent altercation with a car left me shaky and fearful for the day, a reminder of our fragility against metal boxes moving at speed. Research has supported my hunch that Liverpool city centre does not always welcome walkers.[3] There are many active campaigns now to improve access, in Liverpool and across the UK.[4] Part of this must include changing the status of walking if it is valued for its multiple benefits. For too long, walking has been taken for granted or undervalued in planning processes. This reflects the general view that everyday walking is to be tolerated, maybe even pitied. I recall artist Shonagh Short talking about perceptions in her community

where I am from, where I live, if you are walking, it generally means you can't afford the bus and you don't have any friends and family that can give you a lift, neither of which are things to be proud of. Or it's just the quickest and easiest way to get from A to B. It's a non-event. So, we don't go for a walk, we go, we walk. It's an action rather than an event.[5]

Change is happening, and there is a trend towards better pedestrian facilities, but it is hard-won and multifaceted.

147

I spoke to Gloria Gaffney and Don Lee, self-proclaimed 'veteran' campaigners. Gloria has many years of 'pushing herself for people on foot',[6] primarily with Greater Manchester Pedestrian Association (GMPA) and is also a member of Manchester Ramblers and the Peak and Northern Footpath Society (PNFS). Her work with GMPA involves her most; she works to keep people away from traffic danger, noise and pollution. She wants more traffic-free routes. Don has been involved in walking groups for over 50 years and has been part of PNFS since 1962, and the Ramblers from the late 1960s, when he was on their executive. He was a very active local rep within the Open Spaces Society (OSS) from the 1980s until recently. He has now retired from active footpath work 'unless something particularly nasty comes up' and there have been a couple of cases recently. Together, they have a formidable reputation and a long track record of winning battles to maintain or restore Rights of Way. I conducted a walking interview with Gloria and Don for WalkCreate. We met at the People's History Museum and planned to go for a walk along the River Irwell towpath. It was during the interview we spotted the Stopping Up Order for the towpath at Ralli Quays, a serendipitous encounter that led to the campaign discussed in the next chapter.

I had planned to talk to Gloria and Don about how the Covid-19 pandemic changed their walking habits. I also wanted to ask them about their memories of Kinder Trespass celebrations and why they felt footpath preservation was so vital. They told me how fantastic it was to see more people walking in their neighbourhood. They saw a patch of grass become used so often that a desire line became a full path. Gloria also noted how many more dogs she saw and expressed a hope that the walking habit would stick with people. As discussed earlier, the pandemic did see an increase in walking, and many people found this a consolation in difficult times. It was not a panacea though, and the benefits were not experienced equally. For some, quiet streets were a pleasure, but for others, they were a menace.

*   *   *

To enable more equitable walking, change is needed at multiple levels. Many of the barriers transcend gender but a recurrent theme throughout is that women experience public space – and the built environment – differently to men specifically because of their gender. As a psychogeographer, using my body as a tool for understanding place, it is clear to me that the environment can impact behaviour. This is not an argument against the need for behaviour change, or against personal responsibility, but a recognition that people, place, culture and behaviour are mutually constitutive. We make the city, and the city makes us. As geographer Jane Darke says, 'Any settlement is an inscription in space of the social relations in the society that built it ... Our cities are patriarchy written in stone, brick, glass and concrete.' [7] Leslie Kern's *Feminist City*[8] presents a more recent analysis of pertinent research. This is about far more than the phallic symbolism of thrusting towers, the design of buildings have a concrete impact. Safe and accessible walking relies on many kinds of infrastructure.

Maureen Flanagan offers a stark historical example of how barriers manifest in practice by exploring the provision of public toilet facilities for women.[9] In the early twentieth century, there were often violent objections to providing toilets for women because it was assumed they would be used for immoral purposes; it was made clear that decent women should not be in public at all. Today, especially in the context of public service cuts, toilet provision frequently remains inadequate and becomes a very tangible way of exerting control over space and denying the right to be in the city. Public toilet provision clearly still has a gendered dimension. For example, the 'Piss Daleks'[10] introduced in Manchester's Piccadilly Gardens to replace closed civic facilities were urinals designed for able-bodied male anatomies with no additional facilities provided for women, disabled people or anywhere for hygienic disposal of sanitary items. Those constructions have gone, but were not replaced by something more inclusive. WalkCreate heard from many women whose walking was limited by access to toilet facilities. This impacts more than just leisure walking, for example women have been vocal in the truckers' toilet campaign. Furthermore, transphobic rhetoric creates an atmosphere where trans, non-binary and gender non-conforming people feel unable to use facilities and toilet access becomes a way to police gender. As a cis women, I assert strongly I am not threatened by transwomen using the same facilities, and as a disabled person, the majority of toilets I access have always been gender neutral.

Access to safe, hygienic sanitary facilities is vital for everyone's health.[11] Within St George's Hall there is one statue of a woman, Kitty Wilkinson, known as 'Queen of the Washhouse' and 'Saint of the Slums', credited with saving many lives during a cholera epidemic. Her grave is in St James Gardens, every time I pass it fresh flowers have been laid. From where she lies you can see Liverpool's Chalybeate Spring. It was uncovered by quarrymen in 1773, because before the park, before the graveyard, this was where much of the stone that built Liverpool was dug up from. The spring waters were hailed a miracle cure, the minerals imparting healing properties. Unless you boiled it, as the story goes, when it would turn black and sludgy, a warning from the Gods. For many years, the spring was neglected and clogged with bushes. Today, I hear from folk who claim they drink bottles of the stuff, while others caution against its consumption – the water is potentially harmful, contaminated by sewage and other nasty bacteria. I resolve to ask environmental science colleagues about testing the water quality. Meanwhile, launderettes remain an important resource, and can often also function as community hubs. I was reminded of this by artist Merel Smitt when she invited me to join her for a walk as one of her *Launderette Sessions*.[12] One of the cruellest jokes I've seen about social cleansing is the Manchester cocktail bar that you enter though a mock washroom.[13]

Benches are another important aspect of enabling infrastructure. For many people resting regularly is a vital element of going for a walk. There is no one design that works for everyone, so the best option is an array of different seating at various heights, of various sizes, and both with or without armrests. Unfortunately, many local authorities appear to be removing benches or installing more uncomfortable, hostile architecture ostensibly to deter homeless people. An example would be tilted seats or seat breaks to prevent lying down. This does nothing to prevent homelessness, merely stigmatises people and makes life harder for everyone.[14]

Other important enabling infrastructure include adequate, affordable public transport and appropriate lighting. Work carried out by the Women's Design Service (WDS) on *Making Safer Places*[15] frequently found women who never ventured out alone after dark if possible. This means that they do not go out after 4.00 p.m. in the winter. The WDS study found that, in relation to parks and open spaces, women's fears did not seem to be related to actual crime experienced, nor to knowledge or evidence of crime taking place. Fear was related to the possibility of crime occurring, and their wish not to invite it, meaning that women often do not take advantage of green, blue and natural spaces in highly urbanised environments. The perceived risk is too high and often exacerbated by local environmental factors such as poor lighting or difficult relationships with the police.

Lighting and darkness, recur as barriers. As we discussed in Chapter 3 'Eastbourne', women will continue to reclaim the night and walk despite their fears, but this is not an excuse for design inaction. Artists and lighting designers, such as Paula T. Castillo, demonstrate some of the possibilities. She created *The Indoor Sun* for a light festival in Copenhagen, a surreal intervention that sees light pouring onto the street from outside.[16] That may not be a practical solution but it illustrates why a nuanced approach to lighting design is needed. For example, lighting is not just about brightness. The quality and texture of light matters, as do the shadows or stretches of darkness they produce. Environmental factors also need balancing, including light pollution. Nobody wants a city where we can't see the stars.[17]

However, as geographer Gill Valentine[18] stresses, people are a more important factor than physical space when considering fearfulness. She mapped spaces of fear, finding women experience more fear in public space than men, and supporting Jane Jacobs' idea that 'eyes on the street', a busy, activated place, can be a reassurance.[19] Gill Valentine did identify design features such as subways that increase fear and should be eliminated where possible, but you cannot simply design out fear. More pertinently, environmental factors alone will not stop the behaviour that elicits that very legitimate sense of fear. We do know people feel more secure when they feel a sense of belonging. Ingrained social habits, the sometimes subtle, other times brutal, machinations of a patriarchal system mean some women feel they do not belong in public space. A look at the *Everyday Sexism* database, from the campaign founded by Laura Bates[20] illustrates what that system looks like at street level. We must remember the fear women feel is rational and reasonable. It comes from living in a culture saturated with the threat of sexual violence and where popular entertainment too often fetishes dead women.

The WDS closed in 2012 after 25 years due to a lack of funding. In a climate of austerity, their work was deemed low priority, reflecting the flawed idea of a post-feminist milieu where gender has ceased to impact on civil liberties. The Manchester Women's Design Group (MWDG) continued for longer, but is currently dormant. MWDG conducted extensive research into how the urban environment can become more welcoming and inclusive for women, and, ultimately, for everyone. I walked with several members during my thesis research and attended their meetings regularly. MWDG is clear they do not want to essentialise gender but to promote consideration and improved access for all in the built environment. For example, while the majority of childcare remains undertaken by women, they are not, and should not, be the only people responsible for childcare. Design which considers the needs of anyone pushing a buggy or pram[21] will also benefit those with mobility issues, large suitcases or trolleys. MWDG has argued that part of the problem is because women are often poorly represented in the professions most active in urban design, and are generally not in the more senior positions which are most likely to influence policy. Women, therefore, live, work and relax in spaces largely planned and designed by men, where the 'average figure' used to conceptualise space does not reflect reality. MWDG taught me about this when I joined them, and work by women such as by Caroline Criado-Perez[22] show the widespread harm this can cause.

MWDG do assert that, in general, women use the built environment differently from men. The differences can be identified as arising from four major sources: social roles, which mean women are more likely than men to have caring responsibilities and to work part time; low income and relative poverty that impacts disproportionally on women; physical attributes; and culturally accepted norms of where women should and shouldn't go. My Liverpool colleague Catherine

Queen was a contributor to a United Nations report which demonstrates the global need for action to create cities built for women.[23]

Given their desire to create a city more accessible to everyone, it is worth noting why MWDG retained a gendered dimension to their name. This has been discussed within the group several times and consensus is there remains a need to make their feminism explicit. Yasminah Beebeejaun[24] provides evidence that there has been a national shift within public policy to creating spaces that value diversity. This tends to be vague and abstract language that can obscure important struggles. She quotes a member of WDS who highlighted their hard-fought campaigns were now taken for granted: 'we campaigned to get spaces for buggies on buses. We campaigned to get nappy changing facilities in public toilets. Before that women had to change their babies on the floor.' We need to remember these struggles, not least because so many rights are liable to be eroded.

* * *

I think about the difference that a holistic, inclusionary and anticipatory architecture would make to me personally. My office is in a block built in the 1960s. The top floor, the ninth floor, is a common room with a wraparound window. It gives an amazing panoramic view of Liverpool and far beyond. I especially love the view of the Catholic Cathedral. Unfortunately, the lift in the building stops on the eighth floor. I can't verify why, but popular opinion has it that the architects thought a lift shaft would spoil the aesthetics of the roof line. They clearly had no conception that anyone non-ambulatory would ever want to get up there, that anybody who could not walk upstairs would ever belong in a university. Access is always about more than stairs, but they are the amongst the most obvious signs of unwelcome, or ignorance, or disdain.

Joan Rutherford is a retired planner who was chair of MWDG. She is now active in several disabled peoples' organisations and age-friendly initiatives. All centre the importance of accessible infrastructure in their practice. She once took me for a walk along Cown Edge Quarry to admire views across Glossop and some amazing graffiti. Joan was active in the campaign for an accessible Peterloo monument, which was centred around GMCDP (Greater Manchester Coalition of Disabled People). I also got involved, fuelled by sadness and outrage. The Peterloo Massacre Memorial Campaign[25] is a very active and admirable community group which works tirelessly to keep the Peterloo Massacre in the public consciousness. For years, they lobbied for a permanent memorial. I attended several of their events where the names of the dead were recited and found them profoundly moving. Unfortunately, when Manchester City Council commissioned artist Jeremy Deller to design a memorial, nobody thought about access. His vision was of concentric stone circles that could be climbed and used as a rallying point for contemporary rallies. An excellent concept, apart from the steps which meant having a say was only for those who could walk to the top. Disabled people and allies, included Joan and myself, raised concerns as soon as the design was public. After several months we did get to talk to the artist who appeared contrite. Disappointingly though, Deller's memorial went ahead as planned, the only concession being a handrail and a duplication of text at the apex on adjoining ground. Promises were made to fix things later, but nothing has happened.

This is part of the text from a speech I made to an online meeting arranged by Manchester City Council for campaigners on both sides to share their views.

Memorials matter. They are the public face of a city, of its people, its culture. They tell us who and what is worth remembering. They are symbols of power and they have extraordinary resonances. Pay attention to the statues allowed in public space in any city. They tell us who has the privilege to be remembered and celebrated. They are markers of what matters to us and they embody the dominant values of the age and place they were erected ... Peterloo has become a symbol of working-class struggles, of the fight for equality, participation, democracy and inclusion. It is beyond irony that the memorial instead reifies the very opposite. The memorial today embodies inequality, exclusion and segregation.

One reason we were told accommodations would be difficult is its proximity to Manchester Central Convention Centre and so I continued:

If the biggest obstacle to equality is opposition to extending the memorials footprint onto the Manchester Convention City plaza then, respectfully, I ask Manchester City Council to quite literally have a word with itself. Three out of five of the Convention Centre's Directors are based at Manchester Town Hall. We ask for honesty and frankness about who controls public space in Manchester. We ask you to clarify what else you need that space for and whether, actually, it is commercially viable post Covid. An accessible memorial still only uses a fraction of the Plaza ... Disabled people deserve better. Manchester deserves better. Peterloo deserves better.

The memorial reminds us how design and planning are crucial to the experience of cities in general and public space in particular. Conflict can occur when different needs are not considered. Anna Minton[26] makes clear the political and economic policies which shape cities and in particular discusses how many UK city centres have been transformed into somewhere 'clean and safe'. Ostensibly, these are positive attributes to place but Anna documents how they were achieved by ceding the public realm to public-private partnerships and how this fundamentally changes the perception and fabric of a place. In most literature, a 'public space' is defined as a gathering place of the kind that would traditionally be under local authority control, places like squares, plazas and parks. I extend this to pavements and streets because I think these also play a similar role in enabling us to feel part of civic and public life. I have got to know neighbours because we use the same bus stop, or they shared an umbrella with me coming back from the shops.

The changes Anna documents have impacted upon emotions, with new urban barriers paradoxically often generating increased fear and isolation, rather than feelings of safety. For example, those living in a gated community may become paranoid and distressed by imagined or exaggerated threats to their lifestyle, while those outside become resentful and alienated. When I asked Anna how we might change to a more inclusive urban environment, she reinforced the web of power and influence that makes public spaces, telling me 'these places are made for those places [POPS and luxury apartments] they go hand in hand'. Women have been, and continue to be, at the forefront of anti-gentrification campaigns, for example the group of young mothers who formed the Focus E15 campaign for 'Social Housing, not Social Cleansing'. They were galvanised into action when they faced eviction from Newham Council.

I spoke to architect, artist and educator Helen Stratford[27] about what she thought feminist architecture is. She said for her 'Feminism ... extends beyond gender to encompass all

questions of inequality and inequity in public space.' In terms
of architecture, this means

thinking about how people with many different needs; with
many different bodies can be. Some of those needs can
be designed in right from the start rather than becoming
add-ons at the end. For me, it's like, why not? Why wouldn't
you be a feminist? Why wouldn't you strive to make space
for all needs and all peoples?

Helen stresses there is no 'magic formula' for the perfect space,
there are many factors to consider in design. These include
geographical factors, available and appropriate materials,
the intended use and economic considerations, including the
wider political and cultural context. Gender is a part of that.
Helen says, 'the feminist practice of exposing those different
layers, highlights the ways that we might work as architects'.

In her creative work, Helen uses performative actions to
explore power structures in public space. These actions highlight frictions and create a new kind of space while critiquing
what is already there. For Helen, how we perform space, and
the built environment, cannot be separated. Public space is
important to her because they are often 'the most contested
spaces ... but they're also where different possibilities can
happen ... despite those different layers of control, actually,
it's where different kind of possibilities exist continually'.

Recent experiences have made Helen interested in what it
means to lie down in public. Chronic pain made it necessary
for her to lie down on trains, for example, where she was met
with a range of responses from fellow passengers and staff.
She remembers how lying down 'became this non-negotia-
ble demand that my body placed on me ... and even though I
really needed to lie down to counter the pain, I felt ... societal

pressures not to do so ... It was effectively a battle between my body and the space and socially accepted ways of acting.'

Helen explored this experience in her exhibition *Public S/Pacing*[28] at Bloc Projects in Sheffield. The show included flowcharts illustrating how full of labour seemingly 'inactive' lives are alongside exquisitely and symbolically decorated blankets, cushions and beanbags. The work embraced ideas around crip time and collaboration, and Helen explains how she experienced her chronic back pain as a 'continual reassessment and revisiting of spaces and lives, relearning'. She made new spaces of care, asking for, accepting and offering support. Helen worked with many disabled artists, including live artist Rhiannon Armstrong who makes interdisciplinary work and performance but also documents what she calls *Radical Rests*. She describes this as a 'public selfcare system'. She said, 'conceived as a radically accessible version of the viral planking craze, *Radical Rests*[29] concerns itself with making visible the rests that those with disabling conditions have to take in public space.' Raquel Meseguer Zafe's *A Crash Course in Cloudspotting (the subversive act of horizontality)*[30] offers another perspective on this theme. This beautiful, immersive installation incorporates stories from over 250 'resters' living with invisible disabilities or chronic illness. Raquel has also worked with Disorderly Architecture,[31] and Helen is on their advisory board. The collective's starting point is articulated by co-founder Jos Boys as 'Everyone has access needs; it's just that non-disabled people don't recognise that "normal" built surroundings already meet their needs.' Their guide, *Many More Parts than M*[32] is both a superb distillation of creative ways to be inclusive in design and architecture and a vivid illustration of why it matters.

Both Joan and Helen expressed concerns about the lack of depth in training and education about access for architects and planners. It tends to rely on individual educators championing anything beyond legislation, rather than inclusive design being a core part of the curriculum. I remember talking to photographer Saffron Defiance Swansborough about things we wished we had learnt at school. I can only imagine how much better my experience would have been if, for example, the social model of disability was integrated into wider lessons. If the disabling environment was viewed as the problem to fix rather than individual bodies. As people working in education now, we all feel a responsibility for a just and equitable pedagogy that includes discussing issues around social justice and access rights.

*   *   *

Given my previous experiences, it still surprises me I work in a university. As I leave my office, I pass a sculpture by Mitzi Solomon Cunliffe. I love her work in concrete, but perhaps most of all I love the ideas behind her work. Everyone is entitled to beautiful things. She was one of the women celebrated in the *Manchester Modernist Heroines Project* also mentioned in Chapter 3 'Eastbourne'. Cunliffe was an American sculptor and designer who lived in Manchester. She worked on many public art projects, although her most famous work is the Bafta Award Mask. Her mural on Heaton Park Pumping Station links the municipal water supply to its source in the Lake District. What I love most about her is her commitment to the idea that art, and good design, should be for everyone. We all deserve beauty. She wanted her work to be part of the fabric of everyday life, to be 'used, rained on, leaned against, taken for granted', realising her 'life-long dream is of a world where sculpture is produced by the yard in factories and used in buildings as casually as bricks'.[33]

I prefer the walk downhill, and I have time to appreciate the architecture as I think about what to eat. My path intersects with some of Liverpool's Georgian terraces, beautifully proportioned. Then past the bombed-out St Luke's Church, its clock stopped at 3.36. An open air memorial to World War II and the Liverpool Blitz, it's a poignant sight. Today, it's also used as a cultural venue, and there's a soup kitchen that sets up round the side. I'm on Bold Street now, a place where many eras and cultures jostle. There's the radical bookshop News From Nowhere, named after William Morris' *Utopia of Strangers* and Mattas international supermarket which I love. Like Oldham Street in Manchester, it's known for its independent restaurants, cafes and vintage shops, but it does also have chainstores: a Tesco, Greggs and an Oxfam. The provenance of some newer residents is enigmatic. I wonder how many branches Rudy's needs to open before it stops being considered an indie? They originated in Ancoats but their pizza can now be enjoyed in multiple places.

Bold Street has also become renowned in paranormal circles as a hotspot for timeslips. I'm sceptical but love spotting ghost signs, glimpses of the past life of a building where the palimpsest has worn thin or failed to cover up previous uses. Nearby are cobbled streets exemplifying the competing priorities in a streetscape. They are 'authentic', characterful and beloved for their heritage. However, they are a nightmare in terms of access. Wheelchair users can find the jarring painful, and they are serious trip hazards. Cobblestones can also be tricky with a pushchair, buggy or suitcase, or while wearing high heels. They are difficult for blind and visually impaired people to navigate and can also be challenging for some neurodivergent folk.

Over Church Street we are at the border of Liverpool One Shopping Centre. This is one of the first places to really blur the lines between public and private space, and somewhere Anna Minton has written about in detail. It's an arcade with no roof, a mall masquerading as the street. It doesn't feel oppressive, people appropriate it and use it in ways that are not just about commerce. On a sunny Sunday, kids are playing in Chavasse Park, and the skateboarders haven't been moved on yet. However, I have seen protesters and flyer distributors told to leave and disputes over where exactly the boundary is that they cannot cross. I've had similar encounters in Manchester during the Conservative Party Conference. A ring of steel is erected around the convention centre, including ironically, the Peterloo Memorial. We were on a loiter with musician Matt Hill, telling stories of public space and community action and stopped for a song in Barbirolli Square. A policewoman came and told us we could not stand where we were, because it was private, but 'over there' on the same plaza but nearer the Bridgewater Hall was fine. She couldn't show us the line, and none was evident when we returned later.

So much power remains invisible. Nathania Hartley playfully makes some of these forces visible, and audible, through her interventions *Tapping into the City*.[34] Participants are invited to tape coins to their shoes for walks investigating finance and development. Alisa Oleva trains in parkour, and feels the philosophy is very helpful in her walking art because she has learnt to 'treat obstacles as opportunities, if you see something like a wall think of all the endless ways which we can go over, around and under it, not to see it as an end ... see what you can do within that limitation'.[35] She has her own mental map of London parkour locations. In King's Cross, security will arrive in seconds even if you are just rolling around on a bench. In Archway, there is an estate where their work is admired and residents bring them food. This situated knowledge is useful in other ways too. Sometimes Alisa facilitates group walks across London where issues of public/ private land feel very present to her. Sometimes there will be a courtyard or plaza that she wants to share with participants, but knows if they stand and talk they will be confronted by security. They can't dwell there, loitering is forbidden. However, 'the rules change when you are walking'. She notes that merely passing through is confusing to security because fleeting appearances are hard to predict, harder to control or censure.

One of The LRM's most popular games is *CCTV Bingo*. I invented this in 2008 because I wanted a conversation about surveillance. I felt uncomfortable about the proliferation of cameras, but also appreciated for some people they were a positive force. I worked with users of community buildings who were fundraising for CCTV because they wanted to feel more secure. I was also really disturbed by a British Transport Police campaign that suggested merely looking at a camera rendered one suspicious. I felt, still feel, this was fear-mongering and sowing distrust. This was exacerbated by the racialised edge to the posters I saw. Looking, questioning, being curious seem to me to be positive qualities in a citizen. The *CCTV Bingo* game card invites you to find a camera and follow its gaze until you find another. This directs your walk, and while wandering players look for cameras with particular qualities, such as a mobile camera, three cameras facing the same place or somewhere you wish there was a camera.

I'm almost at Albert Dock now. Liverpool lost its UNESCO World Heritage status, but felt it was a worthwhile price to pay for development. Dispersal Orders have been used to stop young people gathering on Mann Island. The Liverpool Slavery Museum is a reminder of how the city's prosperity relied on exploitation and robbing people of their humanity. Slave sales took place opposite the town hall. James Bold, whom Bold Street commemorates, was involved with the slave trade as were many of the prominent gentlemen of his time. Many walking artists work to make visible histories of exploitation. *Ghosts* is an augmented reality tour of Glasgow by Adura Onashile,[36] who worked with the National Theatre of Scotland to create the piece. It is designed to enable mobile phones to show how 'a young man in eighteenth-century Glasgow, leads us on an atmospheric journey of 500+ years of resistance through the streets of the Merchant City down to the River Clyde'. Cathy Turner worked with Keralan artists and students to produce *A Mis-Guide to Kochi*,[37] an invitation to look again, to rub your eyes and see the city that is hiding in plain sight. This reflects Cathy's ongoing interest in interrogating and deconstructing colonial legacies. Amal Paily, Kunji Kuttan Narayanan, Manu Mohan Pallivathukkal, Smija Vijayan and Rithun Manohar Vipin Dhanurdharan (from Kerala), and curator Sumitra Sunder (from Bengaluru) collaborated with Cathy to make *The Mis-Guide to Kochi* which follows other *Mis-Guides* produced with Wrights & Sites.

Wrights & Sites were a huge force in popularising walking art. Formed in 1997, members are Stephen Hodge, Simon Persighetti, Phil Smith and Cathy Turner who say their

> work is focused on peoples' relationships to places, cities, landscape and walking. We employ disrupted walking strategies as tools for playful debate, collaboration, intervention and spatial meaning-making ... Our work, like walking, is intended to be porous; for others to read into it and connect from it and for the specificities and temporalities of sites to fracture, erode and distress it. We have sought to pass on our dramaturgical strategies to others: to audiences, readers, visitors and passersby.[38]

Their first project together, *The Quay Thing* (1998) was a season of site-specific work on Exeter Quayside. Cathy explained creating work related to specific buildings became difficult because the Quayside was the site of proposed city regeneration and therefore a contested space. There was lots of pressure building around what was going to happen to those places, and growing resistance to staged performances in the area. Cathy says,

> we decided the first thing we would do would avoid all that by putting it on a boat ... [we] started to say ... what if performance didn't have to get permission to be in a space but passed through the space and didn't have to be part of this desire to control what a space was but instead would be more responsive to it and to ourselves?[39]

While *The Quay Thing* did eventually include five site-specific performances in and around the quayside, these questions led to a shift in the group's subsequent approach. This change in emphasis let the space be at the centre of everything and decentred the performers. It was 'all about the idea that art and performance could have a light touch, be responsive ... [we] never need to dominate space or put the performer at the centre'. *An Exeter Mis-Guide* (2003) offered prompts for the reader/audience to become the performer and the space. Its success led to a series of other *Mis-Guides*, including *A Mis-Guide to Anywhere* (2006) and *The Architect-Walker: A Mis-Guide* (2018). An early experiment was *Lost Tours 1* (2002), a series of curated walks. These included a 'gender walk', groups of men and women doing the same night walk and comparing experiences afterwards. It was partly due to this experiment that in 2008, Cathy and Dee Heddon began collaborating on their walking women research I discussed in the Introduction.

I asked Cathy about the foundations of her work and she feels the SI have value today in relation to the theory of the spectacle and how walking can identify, displace and disrupt power:

> [its] incredibly pertinent today, very insightful – what [the SI] were doing was closely related to what I feel now is

really important to me about walking … Bringing the focus on the immediacy of experience in a place and allowing your imagination to interact with it … rather than the detachment of a car or a map … technologies that can distance us … separate us from our own experience … they were trying to break through that, that feels like something I really care about.

Cathy also talked about her (non-performative) experiences of walking the South West Coast Path, particularly remembering a moment when she was stressed and the weather was harsh. The walk helped her because 'it just grounds you in your body and in that place, I feel it in a really existential way'.

There's a busker by the waterfront, of course he is playing The Beatles, and excited tourists are taking pictures with him, I notice most don't put any money in his hat and his placard includes a QR code – I hope they are paying him that way. The Beatles' cultural dominance has long overshadowed the richness of musical heritage from a multicultural city. There's a story that records were used as ballast on ships and that was how Northern Soul was born. My Liverpool music scene mostly revolves around AmericanaUK and memories of gigs in flats on Bold Street where the music poured out of the open window and people danced below. The last time I was at the docks was when the Eurovision Village was here, a vision of harmony, glitter and overpriced gin and tonics. I love Eurovision, happily buy into the propaganda. Diving into its archive was one of my pandemic pleasures thanks to a friend who organised virtual contests. It also made me realise what a fairweather fan I am as many were so much more devoted.

I'm walking past the Museum of Liverpool which was used for filming with Jodie Whittaker, the first female *Doctor Who*. Her unveiling as the 13th regeneration moved me to tears as I remember six-year-old me being laughed at for saying I wanted a Tardis when I grew up. I could understand mockery for blurring the lines between fantasy and reality, but this was absolutely because being a Time Lord was not something girls could do. Role models and representation do matter, beneath the surface imaginations bubble and new visions form. If you could go anywhere in time or space where would you choose?

A 2021 *Doctor Who* storyline, *Flux*, partially revolved around the Williamson Tunnels,[40] a labyrinth under the city. I always believed the purpose and extent of the tunnels were a true local mystery. However, a geologist colleague told me recently there is quiet consensus about the 'tunnels'. Investigations suggest they are actually slot quarries that Joseph Williamson excavated on land that he owned. Lots of sandstone was extracted and presumably used as local building stone. Williamson then seems to have worked out that careful use of strong brick arches meant the quarries could be covered over and built on to extract yet more profit from the land. So not really tunnels at all, and in his words, 'Williamson was more than likely a greedy bugger than kind philanthropist.' It's a less palatable and enticing story than the rumours I had heard but it does ring true.

As I head back to Lime Street I walk with the words of Yoko Ono who produced a series of walking instructions (or scripts). Her second 1964 *Map Piece*[41] is deceptively simple and complex. It says 'Draw A Map to Get Lost'. Subverting the conventional function of a map, Ono provides a perfect psychogeographical slogan (although she did not use that term). I'm still thinking about getting lost on the train home. We used *Get Lost!* as the title for a festival in 2008, and saw it as an invitation to play and create new paths. I feel more wary now, and appreciate getting lost is not always desirable. But, when we can, there is something wonderful about the possibilities of drifting through a place we don't know. Dis-orienteering should be a choice available to all. I am filled with rage that so many of us dare not wander in revelry or lose ourselves in exploration because of gendered fear. That fear is not misplaced, I shudder when I recall artist Jane Samuels telling me as a young women in Manchester she was 'cat called 6–7 times a day, assaulted on buses, having to think about where I go at night … it's fundamental to how we live'.

I am not arguing here for recklessness – although adults should be given the dignity of choice to make mistakes and take calculated risks. I am asking for streets that are safe from harassment, cities where infrastructure works and public space is accessible to everyone. It is not an unreasonable demand. Nothing can ever be totally safe but it is absolutely not fair, not right, that so many places are off limits to women. We should be able to reasonably assume we can play without fear we won't make it home unmolested. It shouldn't be hard. The LRM has very few rules. True to our anarchist roots we have faith in people so we just ask this: Take care of yourself, each other and environments we walk through. And don't be an arse.

Back in Manchester, I head out of the station onto Fairfield Street. There's The Star and Garter, a venue that has long been in a precarious position but now seems to have a secure future. It's an important part of the city's intangible heritage for music fans – home of indie discos, alternative club nights, punk gigs and much of my most exuberant and unco-ordinated dancing. Cheers to everyone I met in the queue for the ladies, you are fabulous, and I apologise to everyone who saw my sparkly crochet dress unravel when I spun on, unaware a loose thread was caught on a nail. The venue has been saved by U+I PLC, the developers of the neighbouring land. The area is another that embodies the paradox of contemporary development. It incorporates the first new public park in Manchester for over a century. There's a slide over the newly uncovered Medlock, and accessible play equipment. Brilliant! This is what every woman I spoke to wanted. It isn't utopia though; the park closes at night and often, on a cold morning, there seem to be more staff than visitors. Apartment blocks are being built that will cast shadows, and it's not clear yet how it will impact the surrounding communities in Ardwick Green and beyond. The old abandoned Mayfield Railway Station is part of the development too. Years ago, some loiterers and I wandered up there when the gate was open, disingenuously claiming we didn't know we shouldn't. We stood on the derelict platform and tried to conjure images of trains but instead the Transport Police appeared. The site is now hosting a food hall and the Warehouse Project nightclub. When the regeneration of Mayfield began there was a giant illuminated sign installed on the red brick walls. It read 'everything is connected'.

The LRM returned to Ancoats for a recent First Sunday. It's a landscape utterly transformed since our first loiter there. Crossing the ring road you can still taste the traffic fumes, but there are zebra crossings, cycle paths and a continuous flow of people that the road will not deter. Most of the empty space has gone now and the edgelands have moved further out. A plethora of apartment blocks have transformed the skyline while a marina is home to houseboats, natural wine bars and insta-worthy bakeries. There are tensions with surrounding longer established and less affluent neighbourhoods, and arguments about who should pay for the upkeep of the public realm. However, this evening Ancoats looks smooth, shiny in the twilight. The geese still reign supreme but the retail park has been razed. The women-led 'Trees not Cars' campaigned fiercely for the site to become an actual park, and skateboarders have built their own ramps to ride there. DIY place and community making. The latest plans suggest it will become a 'digital campus' for civil service staff but a portion will be public green space. Time will tell us.

We were joined by artist-architect Dan Dubowitz who asked everyone to bring torches and walking sticks so we could play with darkness and light. The gloaming made windows golden and red bricks muted. We wandered into courtyards, down ginnels, and were invited into a hackspace where 3D printers whirled as visions of a techo-future revealed themselves. Under bridges we peered into crevices, admired cobwebs and studied reflections shimmering in the water. It was dark, proper dark, when we finished and we felt how the beams of light we carried with us changed the atmosphere. Illumination on so many levels, I heard someone whisper 'it's magic, isn't it?'

Dan has conjured other visions of Ancoats too. In 2002, he instigated *The Peeps*. Brass peepholes appeared on the side of buildings throughout the area. An enticement to look closely at what was happening. The viewer was rewarded with phantasmagoria, portals through time and space. Within each peep were illusions, glimpses of industrial archaeology, ghosts or found objects. They included clocking-in machines, brick walls, plants, a steaming toilet. It's bittersweet that *The Peeps* are mostly vanished or broken now, swept away or neglected as regeneration progressed. Dan never shared a definitive map, so there remains the delicious prospect that one day we may discover a forgotten peep hidden in plain site

There's no map of The LRM's walks either, that would be far too didactic for us. Our stories grow and build over time, infiltrating, colouring and constantly rewriting our personal maps of Manchester. It can be hard to find an anchor in a city where the best art is ephemeral, and everything is always changing. Gravity keeps us rooted in the ground we stand in. Looking up, there are still constellations of red crane stars above us. The LRM stood together near the old lockkeeper's cottage and I felt a surge of gratitude for the people I loiter with. I never know who will join me or where we will go but it is always rewarding. Feet, minds and conversation drift and through some alchemical process we move together towards where we need to be.

The script I shared for this chapter reminds me of walking in Sheffield, where signposts point you to the heart of the city. That direction always baffles me. Who knows where the heart is? There has to be more than one; every city is an extravagance of octopuses and we all love different aspects of the same place. In Manchester some of the utility covers proclaim they are 'right at the heart of things' as we trample over them. The heart of my city beats to the rhythm of walking and I realise I can't give a definitive answer to my own prompt. The beat is loud and true, but it is not stationary. The heart of my city is always moving, activated by this peripatetic community of loiterers.

Walking on Together

I crossed the River Irwell. It didn't seem a big deal at the time but I realise now it was. I don't live in Manchester anymore, I live in Salford. I also made the transition from sharing a house to *living with* and that feels even more significant, but like all the biggest decisions I've made there really wasn't a choice. I can angst for hours over minutiae – emails stalled so many times for lack of a comma or a quest for just the right word – but there's a gut feeling that guides the major stuff. I'm typing now from a house filled with boxes I will unpack one day and garish carpets that will be replaced in time. I've learnt that I'm the kind of person who sings loudly (and badly) to themselves all day if there's no one to disturb. I've found peace and comfort in the garden, where we are transforming a pristine manicured lawn into a wildflower meadow. Digging, sowing seeds and planting bulbs is demonstrating faith in the future. We needed help creating the pond, it makes me aware of my bodily limits, but watching tadpoles provoked a childlike glee that took me by surprise. There used to be a steel works nearby and I think about the constants through all the places I have walked: stone, water, metal, trees, love. Solidarities, personal and collective: a state education, the NHS, libraries, public transport, the safety net of benefits like the dole when you needed them, community centres and some very wonderful individuals who took the time to see my potential beneath the chaos and grief. Complex webs of nourishment, support and inspiration that link us all. Walking is never just about feet and floors. In the Introduction, I shared reasons why I loiter, but I should have said 'we' because I was never alone, even when I felt I was. The collective is at the core of The LRM and why I find the Flâneur so objectionable. I can see a lot of the safety nets that caught me disintegrating now, and hatred seems to fester online and in too much political rhetoric. Going outside, walking with, making connections feels more urgent than ever. Small steps perhaps but vital.

The Irwell is a tributary of the River Mersey which flows from Bacup through to Irlam in Salford where it is channelled into the Manchester Ship Canal. For part of its route, it's the boundary between Manchester and Salford, symbiotic cities together yet apart, the river always flowing betwixt and between. Caring for the river is to respect the local environment and recognise how it connects beyond our own patch. Pollutants drift downstream and damage the whole complex ecosystem. For many years, Manchester in particular didn't really value its waterways, seemed surprisingly slow to capitalise on trends for waterside living. The canal still feels like a wild, beautiful secret in some places. I think, perhaps, it's shades of hipcholia. A half-buried shame or sadness about the industrial heritage they remind us of. However, they have been treasured by many and they hold myriad stories. Recently, folk have shared fantastical tales of Messie[1] with me – a glimpse of something strange lurking beneath the surface of the canal. I love a cryptozoological mystery but I suspect it's just debris or litter of some kind, pareidolia maybe induced by intoxicants or an overactive imagination. Perhaps more significantly, what are those Messie rumours really trying to tell us?

On the Manchester side of the Irwell, there's an area rebranded as the Left Bank. It isn't Paris, it's on the edge of Spinningfields, allegedly our Canary Wharf, but it is also home to the People's History Museum. It's a fine place, standing testament to democratic struggles. In 2022, it worked with disabled people to curate *Nothing About Us, Without Us*, a history of Disabled Peoples Activism. I visited several times, feeling inspired and elated but also angry reading about welfare deaths and stigma. So many things we should learn at school, shout about, change. In 2016, the People's History Museum hosted a tenth anniversary exhibition for The LRM called *Loitering With Intent: The Art and Politics of Walking*. It was very strange to see memorabilia from my bedroom in a glass display case. There was an open call for artists and a programme of walks including tours of public toilets, following satellites, exploring histories of LGBTQIA+ activism and tracing flows of money across the city. In the evening, we'd sit by the river eating noodles and chips, and I thought for a bit that might be the end of The LRM. My plan was to go out on a high, before it got boring but loads of people came and re-energised me. I realised there were many loiters left and so much more to discover.

For one of the events, Gloria Gaffney and Don Lee, (the veteran campaigners discussed in Chapter 6 'Liverpool') took us for a walk to see some of the rights of way they saved. They also took us along the towpath opposite the museum. It was a narrow, overgrown, ostensibly neglected trail starting near the tax office at Ralli Quays on the Salford side. It's adjacent to a pub, The Mark Addy, named after a heroic rescuer of drowning people, but the pub was flooded on Boxing Day 2015 and never reopened. Climate crisis means we need to think carefully about what happens to our rivers. I didn't know then how important that towpath would become to me.

Ralli Quays is a singularly unspectacular scrap of land. It's not that enticing, and you need to go down steps to reach it. There's a couple of benches and the boarding point for river tours. It's probably named after the Ralli Brothers, international merchants. Once there was a prison on the neighbouring plot, before that a sacred spring we have yet to pinpoint the precise location of. Nearby is the bridge where the Black Shuck is buried, a monstrous demonic dog who scared decent folk. He is said to be free in 999 years, but no one knows which shore he will alight on.[2]

It turns out many people cherished Ralli Quays. There were plans to build over it, extinguish a path that has endured for 300 years, and divert it through a hotel lobby. Annexation, enclosure, robbing our common treasury. Destroying industrial heritage and access to nature. The fight started slowly, planning committees and council meetings, but soon it escalated. People joined Our Irwell, cleared the path, lobbied hard, fought and won. We saved Ralli Quays. It felt both simultaneously fabulous, massive, and hugely insignificant. We saved a small patch from privatisation. But a win is a win, and what was amazing was the vision this allows to grow. A riverside route from the heart of Salford all the way to the Quays and maybe eventually beyond to Liverpool. We won some hope, forged new alliances and demonstrated our power.

* * *

There are three main areas I think we need to focus on in fights for a better, fairer, more just walking, which also necessitates a more holistic and intersectional definition of access. I believe this has three main components, all entwined and all equally important.[3]

Ownership and legal rights to access

Many of us have been fighting to protect and extend our legal rights to access. This is of real value and importance, but I don't believe this is enough. For meaningful change, we need to demand more than a minimum. We also need to be aware of limiting ourselves to paths defined by others and take a broader of view of what it means to truly occupy space. This might contradict the Ralli Quays example, but we start where we are and if we lost Ralli Quays that vision of a river path would be extinguished.

I believe we need to extend the fight for our rights to roam to urban space, to the streets, parks, greens and public spaces in cities and towns as well as the countryside. These are constantly under threat of annexation and privatisation and many are already lost. This matters because, at its simplest, over 82 per cent of people in the UK live in urban areas.[4] I believe strongly shared spaces play a vital role in building communities as we encounter others and build connections to place. The health and wellbeing benefits of walking and accessing green and blue spaces are well proven. There are also wider environmental benefits to creating more walkable and open cities. My recent experiences with Our Irwell have convinced me of widespread support for a fight to protect and extend urban land rights.

To be clear, we still need to fight for our right to roam in the countryside too but the relationship between rural and urban is both symbiotic and porous. Issues of gentrification, economic segregation and inequality cut across both and are often interrelated, for example around sustainable food, second homes or over tourism. We need to work together for our common cause.

Material issues

Many factors stop people being able to assert their access rights. This is about more than wheelchair ramps, although they are a vital part of the struggle for equality. It includes supportive infrastructures such as benches and shelters, playgrounds, safe and accessible public toilets for all and affordable public transport. It respects all bodies and recognises that we have a collective responsibility to enable and support each other in a spirit of mutual aid and not charity. Making this change means involving, and listening to, everyone who is in a place, or wants to be but can't. It involves improved training and education, and maybe some tactical urbanism too. Move that A-Board that blocks the pavement if you are able to. It also means supporting housing justice campaigns and liberation struggles in any way you can (and yes, sometimes if you can afford it that does mean cash).

Cultural issues

Cultural issues are perhaps the most important, and the most challenging. Many feel certain places are simply 'not for the likes of them' but a true right to roam must be there for everyone. This book focuses on gender and misogyny but we also urgently need to challenge racism, classism, transphobia, ageism, ableism, antisemitism, Islamophobia, other oppressions and their intersections that stop people feeling safe to wander, explore and simply *be* in place. We need to demonstrate practical solidarity and allyship wherever we can, not just sloganeering. This will include listening to others, learning, and we may need to confront difficult truths about our own behaviour, and that of those around us. We also need to acknowledge the labour teaching entails and not take this for granted. We cannot be silent, and this means challenging whenever, wherever and however we can. Active solidarity is needed. Remember the Kinder trespassers were explicit anti-fascists and they can inspire us in this way too. I fully appreciate the need to focus at times on specific aspects or geographical areas, however, I think adopting this model makes it clear we are part of a much bigger struggle. It also illustrates the need to work collectively and co-operatively and to build solidarities across diverse communities and campaign groups. I do not underestimate the challenges involved but believe addressing this would make any movement more powerful and increase the capacity to enable widespread gains and meaningful change.

We will need a variety of tactics, from grassroots conversations to direct action and more conventional campaigning. How can walking contribute? Sometimes I feel utterly powerless and despair at the state of the world. Sometimes a walk is just a walk and there is no pleasure in it – a zombie dashing for a train, or trekking home because I missed the bus. But sometimes stepping out, stepping away, can set a spark alight. Steps flowing, swirling, drifting, your feet are in control and you are wondering where will they take you and who you will encounter. One of the wonderful things about walking is that one never quite knows how it will go. Because of the weather, the environment, other people, we can never be totally sure even if it's a route we have followed thousands of times before. It's never quite finished, never quite static, and we can never quite tell what is around the corner.

The right to loiter

I believe there is a bigger function for streets and pavements in society than their obvious role. The media scaremongers and demonises, but when we are out in the world it is easier to see most people, most of the time, are good. This neighbourliness, this cosmopolitan mingling, does not form strong ties but does help break down some barriers. As culture wars ferment on screen, and artificial intelligence makes it harder to know what fake or good faith is, the ground beneath us remains solid and sound. Being in space, exploring and experiencing shared space, sensing an understanding, makes possible nascent feelings of belonging. The bench, bus stop or corner is an excellent tool for this, it gives the gift of time, of breathing space, of respite to just sit, watch and quietly take your place in the civic sphere. This can enable a sense of being part of somewhere, part of the bustle and life happening all around in the same shared environment.

Conversely, removing those basic rights, that ability to dwell, means alienation, suspicion and fear increases. Differences become potentially catastrophically huge, we don't know who or where we are. Sharing does not always mean agreement. A protest march or demonstration, for example, may cause annoyance, but it still remains central to the democratic health of a city.

Public space is being diminished in multiple ways. To give another example, many cities are banning tents, with policies suggesting homelessness is a 'lifestyle choice' or 'public nuisance' rather than a symptom of austerity and a failure of social policy. Banning tents does not solve any issues but further marginalises homeless people. I would suggest we need to protect and support them and untangle the complex reasons for homelessness. Complex apart from one: financialisation of space has resulted in a lack of affordable housing. More affordable homes is a prerequisite to a solution. If that takes time we need to face up to it. We have a moral obligation to acknowledge we are part of a society that has failed to care for everyone. I hesitate to use the word vulnerable because we are all vulnerable and have the capacity to become more so. It may not be pleasant to pass people sleeping in shop doorways and parks but the bare minimum we can do is acknowledge their humanity. Pushing them to the edges geographically as well as symbolically does not solve the problem, it merely hides our shame. Social cleansing. Policy makers may not like this term, and dress up their choices instead with sham words about safety (whose) cleanliness (key here: what does this mean) and concern (without resources to demonstrate it). Make no mistake, what they are really saying loud and clear is these people are not us, not our responsibility, not our concern. This is a lie.

It can often be instructive to see not who is in place but who isn't. Many hail shared transport spaces a success – this is where cars, bikes and pedestrians use the same space with no demarcations. What is not accounted for in these celebrations are the people who no longer visit because they do not feel safe. You can't count a negative. The same for places bereft of children. Why are they absent? If the environment is not built for you, or sends out signals of ill welcome, you will stay away.

Does this matter? Yes, yes I think it does. Segregation and atomisation are deep problems and we need to do better. We get the cities we deserve. I still get a huge thrill from the opportunities, the crowds, the possibility of new friendships and solidarities. You can find anything and anyone you want in the city. But only if it lets you in in the first place. We need to start from the principle that everyone is valid and worthwhile. This also means confronting racist rhetoric and draconian border controls based on xenophobia and hate.

Investment is needed in infrastructure but the biggest problem is cultural. It is OK not to work all the time, we should not be capitalist drones. We need time to rest, daydream, be unproductive. To loiter should not be a crime, hanging around waiting, recharging are vital and much maligned properties. Give yourself permission to play, to slow down, to really look, listen, feel where you are. Learn about crip time and decolonisation, think about queering space, making new connections and extending invitations beyond what you know. Not everything of value has a monetary price tag and we need to stop operationalising everything. That this can be viewed as radical strikes me as very sad. We all have a right to the city. We can and must shape the place to afford us what we need. Wasteland and underdeveloped plots have a huge role to play here because they are sites of potential, under-prescribed and open for dreams to manifest. This may be walking with a dog, foraging for fruit, enjoying a rendezvous, taking photographs or imagining, but we need the edgelands.

Perhaps the most important thing is walking like you own the space, because actually you do (or should do). You belong here and if that is not obvious then create your own welcoming committee. I recognise the huge privilege I have in being able to exert that desire, and I hope the collective ethos of the loiterers is open enough to provide protection for those feeling insecure. One of the few tenets we have on a First Sunday is that we don't break the law because we recognise not everyone can afford to be detained, whether because of immigration status, caring responsibilities, precarious employment or another reason. We also know how traumatic encounters with the police can be. If it is safe then, walk alone or in companionship as if you own the space already and share it with your equals. Think about how you might extend an invitation.

There can be an astonishing power in a footstep. Desire lines form when an individual (human or not) walks where there is no path and they leave a mark on the landscape. Over time, others follow them and their imprint is deepened. As still others travel along the way, the marks becomes bolder, more defined and then there is a path. We follow in the footsteps of those that came before but still there are a myriad of new paths to create. The eminent geographer and social scientist Doreen Massey taught us that place is a constellation of relationships which ignore lines on maps and transgress across boundaries of all kinds. A carpet spreading beneath us all. Consider where you are now. Do you have a phone or computer nearby? How do you use this to connect with others: loved ones, colleagues, utility call centres. Who are those people connected to? And who made your phone, where did they source the materials and who makes money from you when you use it? We are all enmeshed in webs of relationships but the threads are not all equal in length or strength. Power buzzes along those lines too, vibrating with energy that reflects who is in control. Massey's crucial insight for us is that those connections, those stories, are not finished. We are all still writing them. For Massey, space and place are a tapestry of 'simultaneous stories-so-far'. Implicit in this is the knowledge that this means we can change direction. The ending is not foretold.

* * *

I believe walking in new ways can help us imagine new ways of using, being and connecting with who and where we are. This was illustrated powerfully during the Covid-19 pandemic lockdowns in the UK. This was a deeply distressing and traumatic experience for many and I do not wish to imply walking is a panacea, but for many people walking bought solace, comfort, pleasure and connection. At some points, walking was one of the few permitted activities that enabled people to go outside. Where one could walk was also heavily restricted at times and many were forced to stay in their immediate locality. WalkCreate heard from people who used creative methods to transform their daily walks into something special. This was often spontaneous, vernacular creativity: an instinctive response to make the most of opportunities. Of course that opportunity was not a privilege for everyone. Many key workers continued to work outside their homes and were frequently busier than ever. Other people were shielding and did not venture out for good reasons. As ever, walking embodies and replicates contradictions and inequalities within society. However, for those who could it was often a spark of hope in terrible times. I remember the powerful messages we got in response to our survey questions:[5]

'Walking for me was a way to connect to my community which felt so important during this time. Just seeing other people walking felt reassuring, or seeing pictures people had painted on the pavement outside their house or boxes of books for people to help themselves to, or pebbles people had painted for people to take, was really reassuring and inspiring.'

'It's been one of my lifelines during the pandemic.'

'Walking provides space to connect with others – to smile, speak and share a moment in time. It is a chance to notice, be part of, become curious and celebrate one's surroundings. I have spent years feeling the desire to move away. Walking through Covid-19 taught me to love where I live.'

Clare Qualmann told me: 'I'm more and more motivated and inspired by the quietly radical role of working with people who may or may not perceive what they are doing as art – the vitality of art is its potential to be a catalyst, to shift things for real people in their everyday lives.' This gets to the crux of why walking art, creative walking and psychogeography matter so much to me and why I believe they can be important tools. We need to find ways to imagine better worlds so we can build them, but too often the reality of survival gets in the way. How can we integrate spaces for imagination into everyday life? How can we enchant the mundane, escape the prescribed, even if only for a fleeting moment? How do we resist commodification of everything and turning our lives into a performance? Not everything is for sale. Dériving, drifting or creatively walking together allows us to glimpse other worlds which exist parallel to our own. It may be the world of ghost signs or squirrels or first kisses or secret codes or mushrooms or aliens but it is there, just beyond the surface of our streets, if we take the time to attune our senses to it. We can add new stories to our environment which engender care, connection and communality if we just shift our perspective a little. It should not be radical to walk, talk, dream or take up space in public, but the conditions we strive under can make it so. I'd like to share one final walking prompt, part of a set called *The Metaphysical Treasure Hunt*. It has been collectively written and evolved over time, it carries the DNA of many loiterers within it:

> *The fantastical is breaking through the logic of the city. Look for unicorns, mermaids, fairies, talking mongooses and other impossible creatures. Where do they want to take you? What can they teach you? If you feel enchanted enough ask them to show you a vision of the future (or at least the direction you should head in next).*

An invitation

I know this isn't a recipe to save the world but it is one tool, one among many, which can empower individuals, provoke personal epiphanies and contribute to new visions. Getting to know your place, and your neighbours, feeling the ground and the air and being attentive to the details: these can be positive moves. Loving the local while making bigger connections. Joining invisible dots. An act of hope. It's perhaps telling that Rebecca Solnit has written about hope as well as walking and she says:

> They want you to feel powerless and to surrender and to let them trample everything and you are not going to let them. You are not giving up, and neither am I ... good friends and good principles are worth gathering in. Remember what you love. Remember what loves you. Remember in this tide of hate what love is. The pain you feel is because of what you love.[6]

I love to loiter. We need to fight for the right to loiter for everyone who wants to. We need to take up space, act like we own it, hold onto it and make space for all to join us. Do not get distracted by the traps of authenticity or essentialism. There is never just one way and always be sceptical of definitive views. Expanding to accommodate multiple paths that all converge on one place can be hard work, and requires grace to allow smells and habits that may not be to your taste. It's not romantic but the complexity it allows is so much more beautiful, chaotic, inspiring, brave and real than a singular, top-down 'truth'. Enchanting the everyday act of walking, and opening up our imaginations is one small, positive step we can all take in some way.

The trick we need to learn collectively is holding myriad different cities, different bodies, different paths open all together. We need to defend the right to play, to pause, to dream, to be creative. The ecstasy of a shared moment, enabled by more mundane infrastructures. Feelings that you can pick up and keep warm to help you keep going. They can come in unexpected places. I want a walk where art matters and enchantment is for all. I want to stroll through streets that are thriving, characterful, not sterile or homogeneous. I want to relish a midnight meander home alone and whisper hello to foxes, badgers, owls, bats and frogs. I want to wander in places where the world can be turned upside down and portals can open.

I loiter, and I want us to loiter together, because this space is ours. It belongs to everyone and we are all connected. There is not one true, authentic or right path made by walking, there are many. I began this book with my motivations, but will leave not with an 'I' but a 'we'. The LRM has taught me our power is multiplied when we walk together, collectively, grounded where we are while building wider solidarities. These are the answers given to me over the years to the question 'What is the Collective Noun for a Group of Psychogeographers?':

A walk in progress by The LRM
A situation
A Sinclair
A Self
An Arse
A (de)board
An exploration? An excursion?
A promenade
A reclaiming
A terrain
A wander, a wonder
A meander
An ambling
A line break
A defiance
A trespass
A Kinder, a kinder, a kindling
A kitten
A footstep
A confusion
A São Paulo
A Flâneur, a conundrum cos they are always alone
A co-flâneur
A yard, as in three feet
A ginnel, a twittern, an alley, a jigger
A back passage
A detour
A desire line
An experience
A Luther
A Blissett BLISSit
A drift, a drabble, a dabble
A flân, a stroll, a saunter
A complex

A camber
An ambience
An amble
A shambles
A strew
A loiter
A bimble
A derivation
A diversion
A stomp
A magical mapping
A confluence
A shenanigan
Or us, all of us, any of us, YOU if you fancy it

Consider this an open invitation. I'll be loitering on the next First Sunday of the month and you are very welcome to join me.

Suggested Further Reading

These are some of my favourite books exploring the themes in my work. It is absolutely not a definitive list but I hope it is a useful springboard.

Walking art and creative walking

Helen Billinghurst, Claire Hind and Phil Smith (editors) (2021) *Walking Bodies Papers, Provocations, Actions from Walking's New Movements, the Conference* Triarchy Press.

Elena Biserna (editor) *Walking from Scores* (2022) Les Presses Du Reel.

Dee Heddon, Claire Hind, Maggie O'Neill, Clare Qualmann, Morag Rose, Harry Wilson and Carole Wright (editors) (2022) *The Walkbook: Recipes for Walking and Wellbeing*. Walking Publics/Walking Arts online at https://walk create.gla.ac.uk/the-walkbook/

Claire Hind and Clare Qualmann (editors) (2015) *Ways to Wander* Triarchy.

Alison Lloyd (2021) *Contouring: Women, Walking and Art* Loughborough University Thesis.

Roberta Mock (editor) (2009) *Walking Writing and Performance – Autobiographical Texts by Deirdre Heddon, Carl Lavery and Phil Smith* Intellect Books.

Blake Morris (2019) *Walking Networks: The Development of an Artistic Medium* Rowman and Littlefield International.

Ellen Mueller (2023) *Walking as Artistic Practice* SUNY Press.

Clare Qualmann and Amy Sharrocks (2017) *Study Guide for Women Walking* Live Art Development Agency.

Wrights & Sites (Stephen Hodge, Simon Persighetti, Phil Smith and Cathy Turner) (2006) *A Mis-Guide to Anywhere* Wrights & Sites.

Women walking

Kerri Andrews (2021) *Wanderers: A History of Women Walking* Reaktion Books.

Lauren Elkin (2016) *Flâneuse. Women Walk the City in Paris, New York, Tokyo Venice and London* Chatto & Windus.

Rhiane Fatinikun (2024) *Finding Your Feet: The How To Guide to Hiking and Adventuring* Conway.

Laura Grace Ford (2011) *Savage Messiah* Verso.

Corinne Fowler (2024) *Our Island Stories Ten Walks through Rural Britain and Its Hidden History of Empire* Penguin.

Melissa Harrison (2016) *Rain: Four Walks in English Weather* Faber and Faber.

Jennie Middleton (2021) *The Walkable City Dimensions of Walking and Overlapping Walks of Life* Routledge.

Sonia Overall (2021) *Heavy Time: A Psychogeographers Pilgrimage* Penned in the Margins.

Anita Sethi (2021) *I Belong Here: A Journey along the Backbone of Britain* Bloomsbury.

Rebecca Solnit (2001) *Wanderlust: A History of Walking* Verso.

Elizabeth Wilson (1991) *The Sphinx in the City: Urban Life, the Control of Disorder, and Women* Virago Press.

Louise Ann Wilson (2022) *Sites of Transformation Applied and Socially Engaged Scenography in Rural Landscapes* Methuen Drama.

Psychogeography and the SI

Sadie Plant (1992) *The Most Radical Gesture: The Situationist International in a Postmodern Age* Routledge.

Tina Richardson (editor) (2015) *Walking inside out: Contemporary British Psychogeography* Rowman and Littlefield International.

McKenzie Wark (2024) *Leaving the Twentieth Century: Situationist Revolutions* Verso.

Walking as research method

Charlotte Bates and Alex Rhys-Taylor (editors) (2017) *Walking through Social Research* Routledge.

Maggie O'Neill and Brian Roberts (2019) *Walking Methods: Research on the Move* Routledge.

Stephanie Springgay and Sarah E. Trueman (2018) *Walking Methodologies in a More-than-Human World* Walking Lab.

Foundations

Sara Ahmed (2023) *The Feminist Killjoy Handbook* Allen Lane.

Kimberlé Crenshaw (2019) *On Intersectionality: Essential Writings* New Press.

Paul Dobraszczyk and Sarah Butler (editors) (2020) *Manchester Something Rich and Strange* Manchester University Press.

Nick Hayes and Jon Moses (editors) (2024) *Wild Service Why Nature Needs You* Bloomsbury.

Alison Kafer (2013) *Feminist, Queer, Crip* Indiana University Press.

Louise Kenward (editor) (2023) *Moving Mountains Writing Nature through Disability* Footnote Press.

Leslie Kern (2020) *Feminist City: Claiming Space in a Man-Made World* Verso.

Doreen Massey (2005) *For Space* Sage.

Anna Minton (2009). *Ground Control: Fear and Happiness in the Twenty-First Century City* Penguin Books.

Marion Shoard (1987) *This Land is Our Land* Collins.

Anita Strasser (2020) *Deptford Is Changing: A Creative Exploration of the Impact of Gentrification* www.anitastrasser.com/deptfordischangingbook.htm

Fiona Vera-Gray (2018) *The Right Amount of Panic: How Women Trade Freedom for Safety in Public* Policy Press.

Other media and resources

Jo Norcup (producer) (2016) *Er Outdoors* Three Episodes for Resonance FM.

Michael Umney (producer) devised by Jo Norcup (2018) *The Art of Now: Women Who Walk* A Resonance production for BBC Radio 4 in conjunction with Geography Workshop.

The Loiterers Resistance Movement online www.thelrm.org

The Walking Artists Network online www.walkingartistsnetwork.org

Walk Listen Create online walklistencreate.org

Walking Publics/Walking Art: Walking, Wellbeing and Community during Covid online www.walkcreate.org

Acknowledgements

First, a heartfelt thanks to everyone who has loitered with The LRM through walks, talks, cyberspace, cake maps, games, festivals, TRIPs, exhibitions and so many fabulous First Sundays. Cheers, and here's to many more loiters together.

Thank you so much to the fantastic artists, thinkers and walkers I spoke to for this book: Alisa Oleva, Anna Minton, Cathy Turner, Clare Qualmann, Dee Heddon, Elspeth 'Billie' Penfold, Helen Stratford, Jane Samuels, Jo Norcup, Julie Campbell, Kiera Chapman, Nadia Shaikh, Saffron Defiance Swansborough, Sarah Benjamins and Sonia Overall.

I've been lucky to have crossed paths with many gorgeous people who have supported, inspired and helped me with this book. Special thanks to Adrian Doggett, Alan Smith, Alison Crush, Andrea Tagliatti, Ben Turner, Bethan Evans, Blake Morris, Bren O'Callaghan, Caroline Turner, Ceri Morgan, Corinna Jones, Craig Almond, Dale Meakin, Daisy Porter, Dan Dubowitz, David Dunnico, Dennis Queen, Don Lee, Eddy Hughes, Eleanor Bullen, Emma Curtin, Emma Jackson, Eva Navarro Lopez, Frank Steiber, Gloria Gaffney, Harry Wilson, Hazel Covill, Heena Patel, Helen Billinghurst, Helen Derby, Helen Pendry, Jamie Halliwell, Jenna Ashton, Joan Rutherford, Jo Nightingale, John Piprani, John Purbrick, Jordan Hau, Julian Holloway, Karen Miranda, Kirsty Shires, Konstantina Fotari, Lara Akeju, Lara Nevitt, Lee Johnson, Leo Hollis, Linda Sever, Louise Kenward, Lorenza Casini, Maggie O'Neill, Marie Pattison, Marie Trubic, Martin Green, Matthew Bridson, Matt Hill, Paul Dobraszczyk, Marc Hurt, Mark Robinson, Mark Whitfield, Mike Butler, Mish Green,

Nabeela Ahmed, Natalie Bradbury, Natalie Zazek, Neil King, Neil Walbran, Nick Dunn, Pete Burgess, Peter Overton, Phil Smith, Prenna, Rachel Ravey McLoughlin, Rhiannon Daniels, Sally Hyman, Sarah Irving, Sam Burgum, Sean Fitton, Steve Millington, Serpil Kaya Lindsay, Steven Lindsay, Victoria Henshaw and all at The Basement Collective, BSN, Friends of Library Walk, George Street Community Bookshop, GMCVO, Heyfield Kinder Trespass Group, London Writers Salon, Manchester Women's Design Group, Micro-Climates, The Brownfield Site Immersion Experience, the Fourth World Congress of Psychogeography, Our Irwell, Right To Roam, Walk The Plank and WalkCreate.

Thank you very much to all at Pluto Press, particularly David Castle, for all they have done to make this book happen. Also huge thanks to Rowland Atkinson for sharing his expertise and to Julian Batsleer and Maria Carlos for generous and generative feedback on first drafts.

Thank you to colleagues past and present at the University of Liverpool, the University of Sheffield and Manchester Metropolitan University. I'd also like to thank my students for all their questions and everything they teach me.

Thanks again to the women I walked with for *Women Walking Manchester, Desire Lines through the Original Modern City*.

Thank you to my friends and family. We've taken the scenic route but I am glad to be here now with you all. I hope you know who you are and that I love you.

And, of course, thanks to John Hawes for perfect roast potatoes, pond wading, lifts of many colours and being my true psychogeographical compass.

This book is for Matilda and Nora, Cali and Harper, Amelie, Lyra, Maisie, Molly, Olivia, Peggy, Pearl and Primrose, Marina and Thea: may all your wanders be wonderful.

Notes

First Steps

1. Morag Rose (2008) *Some Psychogeographical Expeditions in Manchester 2005*, self-published and reprinted as part of *A Psychogeographical Treasury*.

Introduction

1. Francesco Careri (2002) *Walkscapes: Walking as an Aesthetic Practice*. Editorial Gustavo Gili.
2. Morag Rose, Dee Heddon, Clare Qualmann, Maggie O'Neill and Harry Wilson (in press) Walking Art. In Peter Adey, Kaya Barry and Weiqiang Lin (editors) *Encyclopaedia of Mobilities* Edward Elgar.
3. Rebecca Solnit (2001) *Wanderlust: A History of Walking* Verso.
4. Kerri Andrews (2021) *Wanderers: A History of Women Walking* Reaktion Books.
5. Morag Rose, Dee Heddon, Matthew Law, Clare Qualmann, Maggie O'Neill and Harry Wilson (2022) *Understanding Walking and Creativity during COVID-19 – Walking Public Report*.
6. Dee Heddon, Maggie O'Neill, Clare Qualmann, Morag Rose and Harry Wilson (2024) Just Walking: Creative Methods towards Pedestrian Equity. In Karen Gray and Victoria Tischler (editors) *Creative Approaches to Wellbeing: The Pandemic and Beyond* Manchester University Press.
7. One of the best introductions to the SI is Simon Sadler (1999) *The Situationist City* MIT Press. More in-depth analysis is offered in the recommended reading list.

8. Guy Debord (1967–1983) *Society of the Spectacle*. Ken Knabb (translator) Rebel Press.

9. Guy Debord (1955) Introduction to a Critique of Urban Geography. In Ken Knabb (editor and translator), *Situationist International Anthology*, revised and expanded edition (2007) Bureau of Public Secrets.

10. Greil Marcus (1999) Heading for the Hills. In *East Bay Express*, February 19, 1999, quoted by Solnit in *Wanderlust*.

11. For these critiques, see, for example, Andrea Gibbons (2015) *Salvaging Situationism Race and Space* in Salvage https://salvage.zone/salvaging-situationism-race-and-space/ (accessed 15 May 2025).

12. For a nuanced account of gender in the SI, see Ruth Baumeister (2020) Gender and Sexuality in the Situationist International. In Alastair Hemmens and Gabriel Zacarias (editors) *The Situationist International: A Critical Handbook* Pluto Press.

13. Greil Marcus (1989) *Lipstick Traces: A Secret History of the 20th Century* Harvard University Press.

14. Manchester Area Psychogeographic (MAP) were most influential to The LRM and there has been some crossover of members. We dematerialised the Beetham Tower partially as a homage to their levitation of Manchester Corn Exchange. Much of their archive is online at www.twentythree.plus.com/MAP/index.html (accessed 31 October 2023).

15. Tina Richardson (editor) (2015) *Walking inside out: Contemporary British Psychogeography*. Rowman and Littlefield International. This is probably my favourite book on contemporary psychogeography, it also includes a more detailed account of the origins and formation of The LRM.

16. Lauren Elkin (2016) *Flâneuse.Women Walk the City in Paris, New York, Tokyo Venice and London* Chatto & Windus.

17. Griselda Pollock (1988) *Vision and Difference: Femininity, Feminism, and Histories of Art* Routledge.

18. Janet Wolff (1985) The Invisible Flâneuse: Women and the Literature of Modernity. *Theory, Culture & Society*, 2(3), 37–46. doi: 10.1177/0263276485002003005.

19. Helen Scalway (2002) *The Contemporary Flâneuse: Exploring Strategies for the Drifter in a Feminine Mode*, www.helenscalway.com/writings/?doing_wp_cron=1506427300.603003025054931640625o.

20. Elizabeth Wilson (1991) *The Sphinx in the City: Urban Life, the Control of Disorder, and Women* Virago Press.

21. Leslie Kern (2020) *Feminist City: Claiming Space in a Man-Made World* Verso.

22. Alison Lloyd (2021) *Contouring: Women, Walking and Art* (2021) Thesis: https://repository.lboro.ac.uk/articles/thesis/Contouring_women_walking_and_art/12490571?file=23168702 (accessed 15 May 2025).

23. Dee Heddon and Cathy Turner (2010) Walking Women: Interviews with Artists on the Move. *Performance Research*, 15(4), 14–22. doi: 10.1080/13528165.2010.539873; Dee Heddon and Cathy Turner (2012) Walking Women: Shifting the Tales and Scales of Mobility Contemporary Theatre. *Review*, 22(2), 224–36. doi: 10.1080/10486801.2012.666741. More recently, see Dee Heddon and Cathy Turner (2025) Walking Women: Site Relational Movements. In Victoria Hunter and Cathy Turner (editors) *The Routledge Companion to Site-Specific Performance*. Routledge.

24. Links to all Jo's audio production and other work can be found here: http://geographyworkshop.com/home/ (accessed 15 May 2025).

25. This is available at the Live Art Development Agency website www.thisisliveart.co.uk/wp-content/uploads/2020/02/WALKING_WOMEN_SRG_FINAL-copy-compressed-1.pdf (accessed 15 May 2025).

26. I take this date from the time we used the name and became more organised. Those first dérives I mention at the beginning of the chapter were ad hoc and not intended to continue.

27. A broad overview of what social centres are, and can be, can be found in Paul Chatterton and Stuart Hodkinson (2007) Why We Need Autonomous Spaces in the Fight against Capitalism. In The Trapese Collective (editors) *Do It Yourself: A Handbook for Changing Our World* Pluto Press.

28. Alex is now a psychology lecturer and author, see Alex J. Bridger (2022) *Psychogeography and Psychology: In and beyond the Discipline* Routledge.

29. Debord was adamant he hated 'isms' and there was no 'situationism'.

30. I have written about how we did this in Morag Rose (2021) Walking Together, Alone during the Pandemic *Geography*, 106(2), 101–4. doi:10.1080/00167487.2021.1919414.

31. Jane Jacobs (1961) *The Death and Life of Great American Cities* Random House.

32. Available to download free from https://etheses.whiterose.ac.uk/id/eprint/19889/.

33. I discuss the impact of this harassment in Morag Rose (2025) 'The City is Not For Us': Ethics, Everyday Sexism and Negotiating Unwanted Encounters during Fieldwork. *Area*, 18 February doi.org/10.1111/area.12997.

34. All outputs from this project, including survey reports, walking art gallery and *The Walkbook* are available to download free here: www.walkcreate.org (accessed 15 May 2025).

35. Solnit's *Wonderlust* gives a comprehensive account of what walking can be, and how its practice has evolved over the years. I recommend her book highly if you are interested in the social and cultural history of walking.

Chapter 1 Manchester

1. This particular script appears on the front of LRM postcards, written in a heart drawn on a map of Bexhill.

2. The Friends of Library Walk website is still at https://friendsoflibrarywalk.wordpress.com/ and you can read an account of what happened in Morag Rose (2020) 'I Am Not a Satnav': Affective Placemaking and Conflict in 'the Ginnel That Roared'. In Cara Courage, Tom Borrup, Maria Rosario Jackson, Kylie Legg, Anita McKeown, Louise Platt and Jason Schupbach (editors) T*he Routledge Handbook of PlaceMaking* Routledge.

3. Doreen Massey (1994) A Global Sense of Place. In *Space, Place and Gender* University of Minnesota Press.

4. Doreen Massey (2005) *For Space*. Sage.

5. Ibid.

6. Massey, A Global Sense of Place.

7. I can't answer that, but it is pertinent to mention that Massey was not convinced by psychogeography, dismissing it as laddish thrills.

8. Get It Done have a specific website about this work here: www.getitdoneart.com/placemakingpiccadilly/ (accessed 15 May 2025).

9. Greater Manchester Law Centre was a leading force in the campaign to stop this and provided a briefing, which can be found here: www.gmlaw.org.uk/wp-content/uploads/2020/03/PSPO-Briefing-Feb-2020.pdf (accessed 15 May 2025).

10. See Anna Minton (2009) *Ground Control Fear and Happiness in the Twenty First Century City* Penguin Books.

11. https://manchestermill.co.uk/the-council-has-closed-off-one-of/ 22.5.22 (accessed 28 January 2025).

12. This argument still rumbles on. More details can be found here: www.manchestereveningnews.co.uk/news/greater-manchester-news/battle-over-250-year-old-17785713 (accessed 15 May 2025).

13. It's actually an advertisement for Converse trainers, but this is not immediately obvious. The moniker I use here was adopted during a Monstrous Manchester walk with geographer Julian Holloway.

14. www.facebook.com/FriendsofLRFS/?locale=en_GB and www.theguardian.com/commentisfree/2015/may/01/london-road-fire-station-activists-citizen-heritage-manchester (accessed 16 May 2025).

15. Jackie Hagan (2015) *Some People Have Too Many Legs* Flapjack Press.

16. This was a Zoom interview. You can find out more about this wonderful project here: www.jane-samuels.com/category/archive/the-abandoned-buildings-project/ (accessed 15 May 2025).

17. Carrie Mott and Susan M. Roberts (2014) Not everyone Has (the) Balls: Urban Exploration and the Persistence of Masculinist Geography. *Antipode*, 46, 229–45. doi: 10.1111/anti.12033 addresses the masculinist tendency in urban exploration.

18. An email conversation, find Julie's music and writing here: https://lonelady.co.uk/ (accessed 15 May 2025).

19. Quote taken from https://lonelady.co.uk/other-work/ (accessed 15 May 2025).

20. See here: www.youtube.com/watch?v=XlK4IoZ8Ndk

21. Jonathan Silver, Desiree Fields, Rich Goulding, Isaac Rose and Siobhan Donnachie (2021) Walking the Financialized City: Confronting Capitalist Urbanization through Mobile Popular Education. *Community Development Journal*, 56(1), January, 161–9. https://doi.org/10.1093/cdj/bsaa044.

22. There is a growing movement doing this work, for example, *The Guardian* (formally *The Manchester Guardian*) Cotton Capital   www.theguardian.com/news/series/cotton-capital/all and the Revealing Histories, Remembering Slavery project http://revealinghistories.org.uk/home.html (both accessed 15 May 2025).

23. See Glen Albrecht, Gina Sartore, Linda Connor, Nick Higginbotham, Sonia Freeman, Brian Kelly, Helen J. Stain, Anna Tonna and Georgia Pollard (2007) Solastalgia: The Distress Caused by Environmental Change. *Australas Psychiatry* 15(Suppl 1), S95–S98. doi: 10.1080/10398560701701288. PMID: 18027145.

24. Sharon Zukin (1995) *The Cultures of Cities* Blackwell.

25. Astra Taylor (director) (2008) *Examined Life*. Contributions from Judith Butler and Sunaura Taylor, Zeitgeist Films

Chapter 2   Ebbw Vale

1. See The EVI. Ebbw Vale Institute www.evi.cymru

2. Ceri Morgan, Emma Bolland, Meg Burkinshaw, Sylvia Crawley, John Mills, Morag Rose, Boo Sujiwaro and Lizzy Trafford (2022) Micro-Climates in NAWE Writing. *Education*, 87 (Summer).

3. Circling Art Project www.circlingartproject.co.uk (accessed 15 May 2025).

4. Louise Ann Wilson *Womens Walks to Remember: With Memory I Was There* www.louiseannwilson.com/work/womens-walks-to-remember (accessed 15 May 2025).

5. Dorothy Wordsworth (1832) Thoughts on My Sick-Bed. Unpublished Rydal Journals.

6. Quote taken from https://walkcreate.gla.ac.uk/portfolio/walks-to-remember-during-a-pandemic-with-memory-i-was-there-louise-ann-wilson/ (accessed 15 May 2025).

7. This was an output from the Arts and Humanities Research Council (AHRC)-funded project *Imagining Better Futures of Health Care for and with People with Energy Limiting Conditions* available here: https://disbeliefdisregard.uk/mish-green/ (accessed 15 May 2025).

8. Jane's work is discussed in Chapters 1 and 5. Here she was talking about her ongoing PhD research at the Centre for Place-Writing at MMU.

9. Their website contains this definition here: www.4wcop.org/. The Fourth World Congress has been held annually since 2015. (accessed 15 May 2025).

10. There is more about Elspeth, who is primarily a ceramicist, here: www.groundworkgallery.com/groundwork-network/elspeth-owen/. The footage of her delivering work to the Victoria and Albert Museum is a joy. (accessed 15 May 2025).

11. Found here: www.walkart.wordpress.com/2013/11/24/hop-skip-jump-by-elspeth-owen/ as part of *On Walking and Art* by Lopez de la Torre (2013) (accessed 15 May 2025).

12. First broadcast on 25 December 2009 www.bbc.co.uk/programmes/boopdjy1.

13. Dee Heddon and Cathy Turner (2010) Walking Women: Interviews with Artists on the Move. *Performance Research*, 15(4), 14–22. doi: 10.1080/13528165.2010.539873.

14. Rosana Cade (2016) The Radical Art of Holding Hands with Strangers. *The Guardian*, 18 August www.theguardian.com/artanddesign/2016/aug/18/radical-art-of-holding-hands-

with-strangers-rosana-cade-walking-holding (accessed 15 May 2025).

15. Lizzie Philps' G.P.S. embroidery can be found here: https://lizziephilps.com/?page_id=350 (accessed 15 May 2025).

16. A Zoom conversation with Elspeth in 2024.

17. Elspeth Billie Penfold *Thread and Word* www.elspeth-penfold.com/home (accessed 15 May 2025).

Chapter 3   Eastbourne

1. Eli Clare (2023) Moving Close to the Ground: A Messy Love Song. In Louise Kenward (editor) *Moving Mountains Writing Nature through Illness and Disability* Footnote Press.

2. 'Weebles wobble but they don't fall down' was the advertising jingle for these toys.

3. I've been refused entry to a bar because the bouncer saw me walk down the street and said I was clearly intoxicated because of my gait. I shared this on social media and realised how common that assumption is after many disabled and neurodivergent people shared their stories of being denied entry to pubs, clubs and other venues.

4. Harriet Larrington-Spencer (2025) Autoethnography of Disability and Active Travel in Greater Manchester: Encountering (Non)citizenship through Access Controls on Traffic-Free Walking, Wheeling and Cycling Paths. *Urban Studies* https://doi.org/10.1177/00420980241311728.

5. This is a well-used SI slogan, and a line in the report by SI member Ivan Chtcheglov (1953) *Formulary for a New Urbanism*.

6. See, for example, Laura Bates (2014) *Everyday Sexism* Simon and Schuster.

7. Liz Kelly (2012) Standing the Test of Time? Reflections on the Concept of the Continuum of Sexual Violence. In Jennifer M. Brown and Sandra Walklate (editors) *Handbook on Sexual Violence* Routledge.

8. See The walkwalkwalk archive here: www.walkwalkwalk.org.uk/ (accessed 22 April 2025).

9. www.jaynejacksonphotography.co.uk/foxstrut (accessed 22 April 2025).

10. Zarina Dolan's work is shared here: https://walkcreate.gla.ac.uk/portfolio/running-shoes-zarina-dolan/ (accessed 22 April 2025).

11. Sara Shaarawi, *Niqabi Ninja* www.independentartsprojects.com/niqabi-ninja/ (accessed 22 April 2025).

12. Eléonore Ozanne www.eleonoreozanne.com/mujeres-paseando-la-ciudad-de-noche/ (accessed 22 April 2025).

13. Women Walk at Midnight www.womenwalkatmidnight.com/ (accessed 22 April 2025).

14. Emma Graham-Harrison (2015) Afghan Artist Dons Armour to Counter Men's Street Harassment. *The Guardian*, 12 March www.theguardian.com/world/2015/mar/12/afghan-artist-armour-street-harassment-walk-kubra-khademi-kabul#:~:text=She%20had%20invited%20almost%20no,rest%20of%20Kabul%20as%20well.%E2%80%9D. More of Kubra Khademi's work can be found here: www.ericmouchet.com/gem/kubra-khademi/?lang=en (both accessed 22 April 2025).

15. Jo Seaman (2015, updated 2022) Museum Crush: The Mystery of the Beachy Head Lady https://museumcrush.org/the-mystery-of-beachy-head-lady-a-roman-african-from-eastbourne/ (accessed 28 May 2025).

16. www.nandomessias.com/sissysprogress.html (accessed 22 April 2025).

17. https://thepansyproject.com/ (accessed 22 April 2025).

18. Paul Harfleet (2024) *Pansy Boy* Barbican Press.

19. Or rather it did. I visited not knowing Myke Beckhurst, the glass artist, planned to retire in October 2024 and the shop closed after 56 years.

20. Hazel, James, Sam, Saffron, Claire and others. Without the now defunct ECAT (Eastbourne College of Art and Technology) I would not be here.

21. Saffron's wonderful photography can be found here: www.instagram.com/cybersaff/?hl=en

22. Fiona Vera-Gray (2018) *The Right Amount of Panic: How Women Trade Freedom for Safety in Public* Policy Press.

23. This was a joint project with Manchester Modernist Society and Shrieking Violet. An account can be found here; Morag Rose and The Modernist Heroines (2022) From an Aviatrix to a Eugenicist: Walking with Manchester's Modernist Heroines. *Gender Place and Culture*. doi: 10.1080/0966369X.2021.1956436; the zine edited by Natalie Bradbury is here: https://issuu.com/natalieroseviolet/docs/manchester_modernist_heroines.

Chapter 4   Stockport, Ashton-under-Lyne and Glossop

1. The project website can be found here: https://eastendjam. wordpress.com/ (accessed 22 April 2025).
2. The project website is here: https://lickablecities.wordpress. com/ Manu J. Brueggemann, Vanessa Thomas and Ding Wang (2018) Lickable cities: Lick Everything in Sight and on Site. In *Extended Abstracts of the 2018 CHI Conference on Human Factors in Computing Systems* https://dl.acm.org/doi/10.1145/3170427.3188399
3. Blake's career has ranged widely, his impressive portfolio is here: https://thisisnotaslog.com (accessed 15 May 2025).
4. Blake Morris and Morag Rose (2019) Pedestrian Provocations: Manifesting an Accessible Future. *Global Performance Studies*, 2(2). doi:10.33303/gpsv2n2a3.
5. Rose, Walking Together.
6. Sonia has written extensively about psychogeography and walking art, her portfolio is here: www.soniaoverall.net/ and Distant Drifts are here: www.soniaoverall.net/events/distance-drifts/ (accessed 22 April 2025).
7. https://bsky.app/profile/soniaoverall.bsky.social   (accessed 22 April 2025).
8. More information can be found here, via co-founder Joe Solo: https://joesolomusic.com/we-shall-overcome/ (accessed 22 April 2025).
9. In Zoom conversation for this work, Dee has written about her *Forty Walks* in Dee Heddon (2012) Turning 40: 40 Turns. Walking & Friendship. *Performance Research*, 17(2), 67–75. doi: 10.1080/13528165.2012.671075; the original blog is here:

https://40walks.wordpress.com/about/ (accessed 22 April 2025).

10. Monique's writing is beautiful and insightful. See here: http://www.moniquebesten.nl/ and https://walkingart.interartive.org/2018/12/monique-besten (accessed 22 April 2025).

11. Jess Allen's work is documented here: https://art-earth.org.uk/jess-allen/ and the quote used is taken from here: https://walkingart.interartive.org/2018/12/Tracktivism (accessed 22 April 2025).

12. Full details here: www.evemosher.com/highwaterline (accessed 22 April 2025).

13. Shonagh's website is here: www.shonaghshort.co.uk/; the quote comes from an interview as part of WalkCreate https://walkcreate.gla.ac.uk/ both (accessed 22 April 2025).

14. The bookshop can be found here: www.georgestreetcommunitybookshop.co.uk/

15. Glossop Heritage Trust archives: 2000 Years of Traffic https://glossopheritage.co.uk/ghtarchive/traffic00/ (accessed 28 May 2025).

16. See https://walklistencreate.org/. The sound walk definition is here: https://walklistencreate.org/about/what-is-a-sound-walk/ (accessed 22 April 2025).

17. In conversation with Dee Heddon for WalkCreate, unpublished interview.

18. Laura Fisher www.laurafisherperformance.com/ (accessed 22 April 2025).

19. In conversation with Dee Heddon for WalkCreate, unpublished interview.

20. Information about all their projects can be found here: https://thedemolitionproject.com/about-the-demolition-project-2/ (accessed 22 April 2025).

21. In conversation with the author. See https://alisaoleva.com/ for more about her practice (accessed 24 April 2025).

22. More about this project can be found here https://alisaoleva.com/walking-home/ (accessed 24 April 2025).

23. *The Walking Library* website is here: https://walkinglibraryproject.wordpress.com/about/ and they have written

about it here: Dee Heddon and Misha Myers (2019). Pedestrian pedagogy: The Walking Library for Women Walking. *Journal of Public Pedagogies*, 4, 108–17.

24. This excellent work can be found here: https://narrowmargins. info/ (accessed 15 May 2025).

25. Phil Smith (2023) *The Silversnake Project* Triarchy Press.

26. The booklet can be freely downloaded here: Emma Jackson (editor) *Writing Walking (One day in Late Spring during a Global Pandemic)* https://research.gold.ac.uk/id/eprint/ 33260/ (accessed 15 May 2025).

27. Louise Rondel and Emma Jackson (2022) Wading into Research: Thinking in and with The River Quaggy. *The Sociological Review*, August https://thesociologicalreview.org/magazine/ august-2022/water/wading-into-research/ You can listen to the landscape here: https://umap.openstreetmap.fr/en/map/ sounding-the-river-quaggy_679393#14/51.4599/0.0087 (accessed 22 April 2025).

28. See www.theguardian.com/commentisfree/2016/mar/10/ swearing-fine-salford-quays and www.theguardian.com/ cities/2019/aug/07/polite-society-why-are-british-cities-ban-ning-swearing-pspo (accessed 16 May 2025).

Chapter 5   Sheffield

1. Victoria Henshaw (2013) *Urban Smellscapes: Understanding and Designing City Smell Environments* Routledge.

2. https://longbarrowpress.com/. Brian documented some of those Lockdown Walks here: https://longbarrowblog. wordpress.com/2021/12/23/last-collection-brian-lewis/ (accessed 15 May 2025).

3. Joanne Lee (2022) 'Sheffield in Virus Time': Forms of Writing, Reading, Living. *Journal of Writing in Creative Practice*, 184–96 http://doi.org/10.1386/jwcp_00038_1.

4. Pete Green (2023) *Sheffield Almanac* Longbarrow Press.

5. Frances Byrnes (2016) The Tragic Story of Sheffield's Park Hill Bridge. *The Observer*, 21 August www.theguardian.com/

global/2016/aug/21/tragic-story-of-sheffield-park-hill-bridge (accessed 15 May 2025).

6. Kristina Rothstein, *In Search of Lost Venues* https://insearch oflostvenues.libsyn.com/about (accessed 15 May 2025).

7. This incident happened in 2000, before I had a mobile phone to take pictures or record evidence.

8. Kiera has a wide body of excellent work here: www.kierachapman.com/ (accessed 15 May 2025).

9. Kiera Chapman, Rowan Jaines, Lulah Ellender and Rebecca Warren (2023) *Natures Calendar: The British Year in 72 Seasons* Granta Books.

10. Imagining Better Futures of Health and Social Care with and for People with Energy Limiting Chronic Illnesses can be found here: https://disbeliefdisregard.uk/. The research team, led by Professor Bethan Evans, has worked on several projects and has included Dr Ana Bê Pereira, Dr Alison Allam, Dr Anna Ruddock, Catherine Hale, Dr Aaliyah Shaikh, Dr China Mills, Dr Stephanie Davis, Hannah Kingston and myself. Project partners are Chronic Illness Inclusion. Healing Justice London and the Lantern Project. We collaborated with the artists Louise Kenward, Julian Gray, Mish Green, Khizra Ahmed and Khairani Barokka (Okka). My thanks to them all.

11. Ellen Samuels (2017) Six Ways of Looking at Crip Time. *Disability Studies Quarterly*, 37(3).

12. Steve Graby (2011) *Wandering Minds: Autism, Psychogeography, Public Space and the ICD.* www.academia.edu/8586966/Wandering_Minds_autism_psychogeography_public_space_and_the_ICD_2011_.

13. Louise's work can be found here: http://louisekenward.com/

14. Louise Kenward, *Moving Mountains*. Chapters cited here are 'Not Healthy, Never Healed' by Isobel Anderson and 'The Clocktower and the Canopy' by Khairani Barokka. It's a superb book.

15. Taken from SEMs website www.semcharity.org.uk/our-history/#:~:text=SEM's%20ethos%20is%20to%20work,a%20good%20quality%20of%20life (accessed 26 April 2025).

16. Morag Rose, Dee Heddon, Matthew Law, Maggie O'Neill, Clare Qualmann and Harry Wilson (2022) *#WalkCreate: Understanding Walking and Creativity during COVID-19*. Public Report Online: https://walkcreate.gla.ac.uk/walkcreate-report/ (accessed 15 May 2025).

17. Rhiane Fatinikun (2024) *Finding Your Feet: The How To Guide to Hiking and Adventuring* Conway.

18. See www.thewanderlustwomen.co.uk/; interview taken from Amelia Hill (2022) The Minority Ethnic Walking Groups Tearing down Barriers. *The Guardian*, 6 July www.theguardian.com/lifeandstyle/2022/jul/06/uk-minority-ethnic-walking-groups-tearing-down-barriers (accessed 15 May 2025).

19. https://collections.vam.ac.uk/item/O107865/pastoral-interludeits-as-if-the-photograph-pollard-ingrid/ (accessed 15 May 2025).

20. Ingrid Pollard Photography: www.ingridpollard.com/ (accessed 15 May 2025).

21. Corinne Fowler (2024) *Our Island Stories Ten Walks through Rural Britain and Its Hidden History of Empire* Penguin. The Colonial Countryside Project can be found here: www.nationaltrust.org.uk/who-we-are/research/colonial-countryside-project (accessed 28 May 2025).

22. Probably the best account of the Kinder Trespass, and certainly the one most often cited by people active in the commemorations is Keith Warrender's 2022 book *Forbidden Kinder: The 1932 Mass Trespass Re-visited* Willow Publishing.

23. More information and research is available here: www.righttoroam.org.uk/ (accessed 15 May 2025).

24. Nick Hayes and Jon Moses (editors) (2024) *Wild Service Why Nature Needs You* Bloomsbury.

25. Jane Samuels, artist. Her website is here: www.jane-samuels.com/ (accessed 15 May 2025).

26. See http://geographyworkshop.com/productions/trees-of-beeston/ (accessed 26 April 2025).

27. https://walkcreate.gla.ac.uk/portfolio/the-meadow-behind-bars-meadowbehindbars-23-march-2020-to-23-march-2021-

alison-lloyd/ More of Alisons work can be found here: www.
alisonlloyd.co.uk/ (accessed 4 May 2025).

28. Anita Strasser *Deptford Is Changing* (2020) self-published see
here: www.anitastrasser.com/deptfordischanging.htm
(accessed 14 May 2025).

29. We wrote about our visit here: Morag Rose and Lorenza Casini
(2022) Fostering a Disconnect: How Verticality Is Hindering
Human Connection with Urban Spaces. *Sociological Review*,
October.

30. A Zoom conversation. The campaign to save Ryebank Fields is
here: www.saveryebankfields.co.uk/.

31. Anita Sethi (2021) *I Belong Here: A Journey along the Backbone
of Britain* Bloomsbury.

32. Doreen Massey, *Landscape/Space/Politics: An Essay*
https://thefutureoflandscape.wordpress.com/wp-content/
uploads/2020/08/landscape-space-politics.pdf (accessed 15
May 2025).

33. https://themanchesterzedders.wordpress.com/ (accessed 15
May 2025).

Chapter 6   Liverpool

1. This script was first used in Blake Morris and Morag Rose
(2019) Pedestrian Provocations: Manifesting an Accessible Future. *Global Performance Studies*, 2(2). doi:10.33303/
gpsv2n2a3.

2. Matthew Worley (2005) *Labour inside the Gate: A History of the
British Labour Party between the Wars* I.B. Tauris.

3. Alex Chapman, Poorva Prabhu and Antony Scott (2023) *Who
Has a Public Right of Way? An Analysis of Provision and Inequity
in England and Wales* New Economics Foundation/Right to
Roam; Jenny Stevenson (2024) *The Most Dangerous Pedestrian
Locations* https://the-compensation-experts.co.uk/no-win-
no-fee-solicitors/guides/dangerous-pedestrian-locations/.

4. For example, the work of Living Streets, the national charity for
everyday walking www.livingstreets.org.uk/; Paths for All in
Scotland www.pathsforall.org.uk/; for a regional example, see

Walk Ride GM www.walkridegm.org.uk/ (all accessed May 15 2025).

5. Unpublished interview for WalkCreate.

6. Unpublished interview for WalkCreate.

7. Jane Darke (1996) The Man Shaped City. in Chris Booth Jane Darke and Sue Yeandle (editors) *Changing Places Women's Lives in the City* Sage.

8. Kern *Feminist City*.

9. Maureen Flanagan (2014), Private Needs, Public Space: Public Toilets Provision in the Anglo-Atlantic Patriarchal City: London, Dublin, Toronto and Chicago. *Urban History*, 41(2), 265–90. doi:10.1017/S0963926813000266.

10. I am indebted to the brilliant David Dunnico for this phrase. David Dunnico (2014) Inconvenienced: How the Cuts Have Hit Public Toilets www.redpepper.org.uk/inconvenienced-how-thecuts-have-hit-public-toilets/. David also took the image shared on the latest iteration of the CCTV Bingo Card.

11. Jen Slater and Charlotte Jones (2017) *Around the Toilet: A Research Project Report about What Makes a Safe and Accessible Toilet Space* provides an excellent summary of what this could look like https://aroundthetoilet.wordpress.com/.

12. https://merelsmitt.nl/the-launderette-sessions/ (accessed 15 May 2025).

13. Dale Lately (2017). Pawn Shops and Poverty Chic: How Working-Class Life Was Colonised. *The Guardian*, 17 May www.theguardian.com/cities/2017/may/02/poverty-chicworking-class-urban-life-colonised.

14. See Radhika Bynon and Clare Rishbeth (2016) *Benches for Everyone: Solitude in Public, Sociability for Free* The Young Foundation for more about why benches matter.

15. Womens Design Service (1998) *Making Safer Places.* Their web archive is here: www.wds.org.uk/pub_current.html (accessed 15 May 2025).

16. www.paulacastillot.com/about (accessed 15 May 2025).

17. See Nick Dunn (2025) *Dark Futures, a Manifesto for the Nocturnal City* Zero Books for ideas.

18. Gill Valentine (1989) The Geography of Women's Fear. *Area*, 21(4), 385–90.
19. Jacobs *The Death and Life of Great American Cities*.
20. Bates *Everyday Sexism*.
21. For more on the challenges of walking with a buggy, see the excellent work by Louise Platt, including her 2023 *Walking with Infants: A Manifesto for Walking Mums* Leisure Studies Association/MMU walkingwithinfants.blogspot.com/. Clare Qualmanns' Perambulator Series (2012–23) is another aspect of her walking art practice.
22. Caroline Criado Perez (2019) *Invisible Women: Exposing Data Bias in a World Designed for Men* Chatto and Windus.
23. Sara Candiracci and Kim Power (2022) *Cities Alive: Designing Cities That Work for Women* Arup, UNDP and University of Liverpool www.undp.org/publications/cities-alive-designing-cities-work-women (accessed 15 May 2025).
24. Yasminah Beebeejaun (2017) Gender, Urban Space, and the Right to Everyday Life. *Journal of Urban Affairs*, 39(3), 323–34. doi: 10.1080/07352166.2016.1255526.
25. The Peterloo Memorial Campaign: www.peterloomassacre.org/ (accessed 15 May 2025).
26. In conversation, also documented in her works *Ground Control* and (2017) *Big Capital: Who Is London for?* Penguin Books.
27. In Zoom conversation. Helen's work is collected here: www.helenstratford.co.uk/
28. Debbie Rolls writes about this exhibition here: https://corridor8.co.uk/article/helen-stratford-public-s-pacing/ (accessed 15 May 2025).
29. Rhiannon Armstrong Radical Rests can be found here https://www.rhiannonarmstrong.net/projects/radicalrests/ (accessed 16 June 2025)
30. Raquel Meseguer Zafe A Crash Course in Cloudspotting https://uncharteredcollective.com/performance.
31. https://disordinaryarchitecture.co.uk/start-learning/doing-disability-differently (accessed 15 May 2025).
32. The Disorderly Architecture Project's brilliant book, *Many More Parts Than M: Reimagining Disability, Access and Inclusion*

beyond Compliance is available here: https://disordinary architecture.co.uk/start-learning/many-more-parts (accessed 15 May 2025).

33. Quoted in Lynn Pearson (2007) Roughcast Textures with Cosmic Overtones: A Survey of British Murals, 1945–80. *The Journal of the Decorative Arts Society 1850–the Present*, 31, 116–13.

34. Nathania Hartley, *Tapping into the City* https://nathaniahartley. com/projects/tapping-city (accessed 28 May 2025).

35. In conversation, 2025.

36. www.nationaltheatrescotland.com/past-performances/ghosts (accessed 22 April 2025).

37. https://expandeddramaturgies.com/a-mis-guide-to-kochi/ (accessed 22 April 2025).

38. www.mis-guide.com/ (accessed 22 April 2025).

39. In conversation.

40. www.williamsontunnels.co.uk/ (accessed 15 May 2025).

41. Yoko Ono *Map Piece* (1962–1964). See Yoko Ono (1964, 1970, 2000) *Grapefruit A Book of Instructions and Drawings* Simon and Schuster.

Walking on Together

1. My thanks to Craig Almond who named the enigma during a Twitter conversation.

2. The folklore of these waterways is rich and revealing. 'Manchester's Ophelia' Lavinia Robinson is particularly pertinent to this book, see Morag Rose (2019) There's Something in The Water! A Psychogeographical Exploration of Manchester's Waterways in Karl Bell (editor) *Supernatural Cities: Enchantment, Anxiety and Spectrality* Boydell and Brewer.

3. This section was first shared at events discussing the Kinder 91 events.

4. Department for Environment Food and Rural Affairs (2021) *Statistical Digest of Rural England*.

5. These quotes are all from the public survey, available here: www.walkcreate.org (accessed 15 May 2025).

6. This was a public Facebook post by Rebecca Solnit on 6 November 2024 and can be accessed here: www.facebook. com/story.php?story_fbid=10161821712395552&id=5598 35551&mibextid=CTbP7E&rdid=NoG9VuwPTvJQ6to9# (accessed 15 May 2025).

Index

The Pluto Press Newsletter

Hello friend of Pluto!

Want to stay on top of the best radical books we publish?

Then sign up to be the first to hear about our new books, as well as special events, podcasts and videos.

You'll also get 50% off your first order with us when you sign up.

Come and join us!

Go to bit.ly/PlutoNewsletter